BOEING B-17
FLYING
FORTRESS

1935 onwards (all marks)

Published in North America in 2011 by Zenith Press,
an imprint of MBI Publishing Company,
400 1st Avenue North, Suite 300, Minneapolis,
MN 55401 USA, by arrangement with
Haynes Publishing.

Zenith Press titles are also available at discounts
in bulk quantity for industrial or sales-promotional
use. For details write to Special Sales Manager at
MBI Publishing Company, 400 1st Avenue North,
Minneapolis, MN 55401 USA.

To find out more about our books, join us online at
www.zenithpress.com. or www.qbookshop.com.

ISBN-13: 978-0-7603-4077-6
ISBN-10: 0-7603-4077-3

Printed in the USA by Odcombe Press LP,
1299 Bridgestone Parkway, LaVergne, TN 37086.

COVER CUTAWAY:
Boeing B-17G Flying Fortress. *(Mike Badrocke)*

BOEING B-17
FLYING FORTRESS

1935 onwards (all marks)

Owners' Workshop Manual

An insight into owning, restoring, servicing and flying
America's legendary World War II bomber

Graeme Douglas

The Collings Foundation's airworthy
B-17G, 44-83575, *Nine O Nine*.
(Patrick Bunce via Keith Wilson)

Contents

◁——(●)——▷

Acknowledgements

A project of this size requires the help and support of many people, who have contributed in different ways to make this book possible. The author would like to offer his grateful thanks to the following:

Association Forteresse Toujours Volante – Michel Bézy, a true B-17 enthusiast, who patiently answered many questions and assisted with photographs. Also, to the entire crew of *The Pink Lady*, who were all welcoming and helpful during the author's visits to France.

Deltair Airmotive Limited – Chris Adams, who gave up an afternoon of his valuable time to enable photographs to be taken of a stripped Wright Cyclone engine.

Haynes Publishing – Jonathan Falconer, for assistance and encouragement to the author.

Imperial War Museum, Duxford – Andy Robinson for his valued help during photo-shoots inside *Mary Alice* and for supplying photographs; Jessica Jeske and John Delaney for arranging the logistics of the photography at Duxford.

Many individuals contributed advice, information, assistance and photographs. Special thanks go to Peter Brown, for many years the chief engineer on *Sally B*, whose knowledge is second to none on all things technical, and who has been an endless source of information and encouragement; Steve Carter for supplying many of the maintenance photographs; thanks also go to Mike Stapley for maintenance information.

Grateful thanks to Dave Littleton for allowing the author free access to his superb B-17 cockpit to take photographs, and for sharing his considerable knowledge of all things B-17; Frank Talbot for his assistance during the author's visits to France – from offering accommodation, to research, translation, public relations and taking many valuable photographs. His assistance has been invaluable; the re-enactors, who went to so much effort for the mission photos – Tom Achard, Dave Pratt, Laurence White, Andy Burley, Tony Hollies, Tim Parker and Dave Spratt; author and historian Scott A. Thompson for information and for allowing the use of a photograph of *Mary Alice*; Gerard Boymans, Joe Rimensberger, Patrick Bunce and SFB Photographic/Keith Wilson for their superb air-to-air photographs. Other photographs were supplied by Pieter Kroon, Bruce Orriss, the Imperial War Museum, London, and the RAF Museum, London.

The author is indebted to pilot Darrell Blizzard for allowing him to use his account of wartime flying missions in the B-17 and for the use of his crew photograph; thanks also to the 351st BG historian Ken Harbour for information on the history of *The Pink Lady*.

Many years ago a young volunteer restorer at Duxford became hooked on the B-17. Thank you, Ted Hagger.

Lastly, a special note of thanks goes to Marion Douglas for her proofreading, translation, grammar corrections, and suggestions for punctuation. Above all, thanks for her great patience during this time! I dedicate this book to her.

Author's note

During the writing of this manual one of the featured aircraft, *The Pink Lady,* has been retired. She is now based at Cerny airfield in France and on public display. As yet, her future

BELOW Boeing Flying Fortress manufacturer's plate. *(Graeme Douglas)*

is unknown. There is talk that a new generation of engineers will help to maintain her and – hopefully – return her to the air. With the wealth of experience that exists within the current team, it must be hoped the knowledge is passed on to those who follow so that *The Pink Lady* can once again take to the air.

The Pink Lady and Duxford's *Mary Alice* were chosen as twin centrepieces for this book for different reasons. *The Pink Lady* is, at the time of writing (2010), the only B-17 still potentially airworthy which is a genuine combat veteran, having flown six missions during April 1945. As such the aircraft is a link with the past and a genuine survivor when so many of the world's preserved B-17s, both airworthy and static examples, saw no combat service despite the illusion given by their authentic-looking colour schemes. Internally *The Pink Lady* has been modified from her original military configuration, having been altered over the years for several different roles. Of all the surviving B-17s there are few, if any, which have not been modified significantly by the removal of military armament and equipment not required for peacetime civilian roles. In its place was substituted camera equipment, special search radars or water-bombing equipment, depending on post-war operators' requirements.

Some restored B-17s have undergone an internal refit back to a military specification. Of these, some examples are more thoroughly restored than others. It has been said that *Mary Alice* at the Imperial War Museum, Duxford, is one of the most thorough restorations in existence. For this reason, and for the fact that the author spent 17 years as a volunteer restorer on this particular aircraft, it has been chosen for this book. It represents a B-17 that has been fully restored to an original USAAF military specification, demonstrating the features and equipment found on board a B-17G during the latter part of the Second World War. Most of the internal photographs reproduced in this manual are from this aircraft.

The author has also made use of photographs gathered while working as a member of the ground crew on the UK's airworthy B-17, *Sally B*.

For the close-up photographs of the engine start procedures, an authentic replica cockpit built by David Littleton was used. This gave easy access for the camera when engaged in close-up photography.

Graeme Douglas
Essex, September 2010

ABOVE When construction ceased in 1945, 12,731 Boeing B-17 Flying Fortresses had been built. To mark the completion of the 5,000th B-17 built at Seattle, Boeing invited each member of its workforce to put their signature on the aircraft. *(Boeing)*

'Without the B-17, we might have
lost the war.'
General Carl A. Spaatz
Commander of US Strategic Air Forces in Europe 1944–45

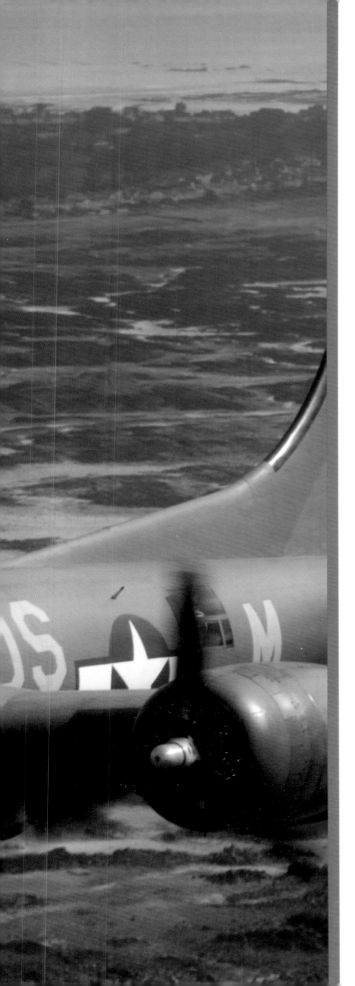

Introduction

As one of the most potent symbols of American military strength in the Second World War, the Boeing B-17 Flying Fortress has achieved something of an iconic status. To the public in wartime Britain watching the massed formations of B-17s overhead, it meant their nation was no longer alone. To the American crews who flew the Fort it represented a sturdy, dependable aircraft capable of absorbing massive damage and still returning them home.

OPPOSITE B-17 Flying Fortress *The Pink Lady*. *(Gerard Boymans)*

For today's aviation enthusiasts the sight and sound of one of the few remaining airworthy Flying Fortresses still creates an emotional response as they hear the liquid rumble of the engines and watch the graceful curves of 'The Queen of the Skies' as she sweeps majestically along the display line at an air show.

Blessed with an evocative name that caught the public imagination of the Forties, the legendary Fortress still has an appeal that is forever associated with the friendly invasion of so many Americans to a small corner of England, and with the enduring music of Glenn Miller.

The US Army Air Forces in Europe took the war to the enemy by day. Their war was not fought in the stealth of night like RAF Bomber Command, but in the punishing daylight skies where combatants could see and kill one another readily. It can be easily forgotten why the Fortress was conceived, what it was built to do, and how much death and destruction it caused. In the cause of freedom so much life was lost – but to remember those lives today is the reason above all that there are still Fortresses flying.

BELOW The elegant proportions of the Flying Fortress earned it the epithet 'The Queen of the Skies'. (Joe Rimensberger)

FORTRESS SURVIVOR – *THE PINK LADY*

The Pink Lady, 44-8846, is one of the few surviving B-17s. She is unique. The aircraft, now based in France, is the only B-17 still potentially airworthy which saw combat service, having flown six missions at the end of the Second World War. To date she has completed more than 10,000 flying hours and it is planned that she will undergo long-term restoration with the aim of her flying again to delight air show crowds and commemorate the past.

The Pink Lady has enjoyed a long and varied history. She has been based in many countries and has flown in a variety of roles and under several different identities, as well as undergoing many technical modifications over the years. The story of this remarkable survivor began in December 1944, with the prospect of the war in Europe drawing to a close. The production of B-17Gs at the Lockheed Vega factory in California was beginning to wind down. At its peak, in the spring of 1944, the Boeing, Vega and Douglas factories were producing a combined total of more than sixteen new aircraft a day. Now, the Burbank factory rolled out no more than two or three finished bombers each day. The factory would be the last of the three to close, in August 1945, after completing 2,250 aircraft.

One of those completed at the end of the last full year of the war was a B-17G allocated USAAF serial number 44-8846 (referred to simply as '846'). The natural metal-finish Fortress was part of production block B-17G-85-VE, with manufacturer's serial number 8246. At this late stage in the war, B-17Gs were left unpainted to reduce time spent in the factory applying paint. It was also deemed unnecessary to camouflage the bombers because the mass formations flying over hostile territory were often clearly visible from the ground, given away by the streaming contrails from their engine exhausts.

On 17 January 1945, 846 was flown to the United Air Lines modification centre at Cheyenne, Wyoming, for final fitting out for its flight to England to join the US Eighth Air Force. Here, the aircraft was marked out for special modifications which would give it a different

role to the standard bomber at the time. This work entailed installing a special radar unit in the position previously occupied by the ball turret. Commonly known as 'Mickey', this installation gave the aircraft the ability to bomb through cloud cover and, as such, 846 would

have the role of a squadron lead aircraft. With the modification work completed, the aircraft was flown to the UK and in March 1945 she was assigned to the 511th Bomb Squadron (BS) of the 351st Bomb Group (BG) based at Polebrook, Northamptonshire. Here, the

ABOVE *The Pink Lady* – also referred to as '846' – flying near La Ferté-Alais, south of Paris, where for many years she has displayed at the annual air show. *(Joe Rimensberger)*

LEFT Streaming exhaust contrails gave away the presence of high-flying bombers like the B-17 to observers on the ground. *(USAF)*

ABOVE A dedicated team of ex-Air Inter engineers have worked to keep 846 flying for more than 25 years. Here, routine maintenance is taking place on the engines. *(Graeme Douglas)*

RIGHT The bombardier's position in the nose of *The Pink Lady* offers superb views. Notice that the Norden bomb sight has been covered over, as it would have been during the war years, to prevent photographs of it being taken that could fall into enemy hands (although its technical details were already known to the Germans). *(Franck Talbot)*

markings of the 351st BG, consisting of a white letter 'J' in a black triangle with a diagonal red stripe, were applied to the tail fin. The squadron codes of the 511th BS, the letters 'DS', were applied to the fuselage sides with the aircraft's individual identification letter, 'M'.

While still based at Polebrook, 846 acquired the name of *Half Pint* along with a pin-up in a green swimsuit that was painted on the nose of the aircraft. Between 8 and 20 April the aircraft flew six combat missions to France and Germany, which makes it a unique aircraft as it is the only B-17 still potentially airworthy which saw combat service during the Second World War. While most of the 351st BG aircraft returned after the war to the United States for eventual scrapping, 846 was saved from this fate by being reassigned to the 305th BG at Chelveston, Northamptonshire, in Project 'Casey Jones', an operation to photograph and map Germany from the air.

The aircraft regularly changed operating bases across Europe. Assigned to the 45th Reconnaissance Squadron, 846 was based initially in Belgium and then Lechfeld in Bavaria. In May 1949, 846 was converted to a photo-reconnaissance RB-17G specification, and was based first at Fürstenfeldbrück airfield from where it carried out secret flights mapping the borders with the Soviet Union and in the Berlin corridor area. This was an ironic twist of fate because in April 1945, 846 had attacked this very airfield while flying with the 351st BG from Polebrook. In August 1950 the aircraft was moved to Wiesbaden in Germany. It continued to serve in the photo-reconnaissance role until 1953 and then returned to the US.

In November 1954 the aircraft was retired from the USAF and put up for disposal. Its fate lay in the balance. 846 was saved from the scrap man's blowtorch only because it was bought by the Institut Geographique National (IGN), the French Geographical Institute, and flown to France in December of that year where it was given the civil registration F-BGSP. In this guise it operated from the IGN base at Creil, near Paris, alongside other B-17s engaged on high-altitude mapping and survey work.

BELOW The distinctive nose art of *The Pink Lady* is shown to good effect here. The design pays homage to the many B-17s which sported such artwork during the Second World War. *(Graeme Douglas)*

ABOVE This profile study shows the large tail and dorsal fin which characterised the B-17E series onwards. *(Gerard Boymans)*

LEFT Known as *Lucky Lady* at the time, 846 is captured at the moment of touchdown at St-Yan, France, in 1987. *(Joe Rimensberger)*

In this role it flew until 1979 with the IGN, accumulating almost 9,500 hours of flight-time in photo-mapping roles all around the world.

There then followed a period of uncertainty for the aircraft. It was placed into storage and still maintained, but with no immediate plans to fly again. Thanks to a request from the French defence ministry for 'F-SP' to fly at the annual 14 July Bastille Day celebration in 1984, the aircraft was returned to flight. Following this successful return to the air,

RIGHT *Lucky Lady* on the ground at St-Yan. *(Joe Rimensberger)*

CENTRE In natural metal finish, the B-17 contrasts with the green fields of France. *(Joe Rimensberger)*

BOTTOM Pictured at La Ferté-Alais in 1983, 44-8846 still carries the IGN civil registration and spurious Eighth Air Force markings. *(Joe Rimensberger)*

F-SP was flown by an IGN crew at a number of air shows. Eventually, to ensure continued public appearances of the historic bomber and to safeguard its flying future, a deal was struck between the IGN and the *Association Forteresse Toujours Volante* (FTV). FTV purchased the aircraft with an agreement that the IGN would train a pilot, André Dominé, and an engineer, Patrick Blanquart, both of whom had extensive experience on heavy aircraft. This association, in partnership with the well-known French warbird organisation Amicale Jean-Baptiste Salis, established a long-term strategy to preserve the aircraft and keep it flying for a period initially intended to be five years. In August 1987 the B-17 flew its first air show under the FTV banner, initially flying as *Lucky Lady* and now carrying the civil registration of F-AZDX, operating from Orly airport.

When André Dominé began to operate the old bomber from Orly airport, he started to arouse the interest of a number of ex-French Air Force engineers who were working for the French airline Air Inter. These engineers, with their historical links to piston-engine aircraft, were drawn to the B-17 and volunteered to help maintain and service the aircraft. Most of them are still working on *The Pink Lady* more than 20 years later. The group are a tight-knit community who share a similar working background and now in their retirement they share a common purpose in maintaining this historic aircraft.

The aircraft's 'working' years were over and now began a more relaxed period of appearances at air shows, as well as in films.

In the summer of 1989, the feature film *Memphis Belle* went into production in England.

ABOVE The B-17's good handling characteristics make it well suited as a display aircraft. *(Joe Rimensberger)*

RIGHT By 1989, 846 was painted in the colourful markings of a 486th Bomb Group (BG) aircraft. Here she is pictured awaiting take-off at an air show in Friedrichshafen, Germany. *(Joe Rimensberger)*

The film gives a fictional account of the true story of the first B-17 based with the Eighth Air Force in England to complete 25 missions. *Lucky Lady* flew to Duxford to take part along with four other B-17s that had been gathered together for the filming. Required to depict an earlier B-17F, *Lucky Lady* was painted in olive drab upper surfaces with grey undersides and fitted with mock-up ball and upper turrets and twin tail guns. The squadron codes 'DF' were applied to the fuselage sides, together with individual aircraft identification letters, which were changed regularly throughout the filming to create the impression of many more aircraft. *Lucky Lady* assumed a dual identity during the production: on the left-hand side the name *Mother and Country* was applied, and on the right, the name of *The Pink Lady*.

When the filming ended, the aircraft retained the twin identities worn on camera, but she became known universally as *The Pink Lady*. Today, as well as this nose art, the aircraft carries the bomb group and squadron markings that were applied when she first entered service with the 351st BG of the USAAF way back in 1945.

Over the years, *The Pink Lady* has made air show appearances across Europe – in Belgium, Switzerland, Germany, Austria, Great Britain and all over France. In 2009 the aircraft flew to the Czech Republic to take part in the making of *Red Tails,* a film about African-American Tuskegee airmen in the Second World War.

In the late 1990s the stress of nearly 60 years of flying started to be noticed in *The Pink Lady* and other veteran B-17s. Operators in the United States became aware that cracks and corrosion were developing at the attachment

ABOVE **For the film *Memphis Belle* (Michael Caton-Jones's 1990 remake of the classic 1943 documentary of the same name by William Wyler), 846 adopted olive drab and grey camouflage paint scheme depicting the B-17 *Mother and Country*.** *(Joe Rimensberger)*

ABOVE During the making of *Memphis Belle,* 846 was one of five B-17s that took part in filming at Binbrook and Duxford. Their markings depicted aircraft from the 91st BG at Bassingbourn. *(Andy Robinson)*

RIGHT As *Mother and Country*, 846 takes off from Duxford on a filming sortie. *(Andy Robinson)*

points where the wings joined the fuselage. Although this did not affect all aircraft an investigation was launched, which led to an Airworthiness Directive (AD) being issued by the Federal Aviation Authority (FAA) in 2001. This is a legal document which requires owners and operators to comply with the instructions to investigate and rectify these problems. Although issued in the US, the AD applied worldwide to all B-17 operators if they wished to continue flying their aircraft.

There were two ways to comply with the AD. The first was to carry out regular inspections of a number of key wing retaining bolts, to check the integrity of both the bolts and fitting hole. The other method entailed removing the wings from the fuselage and replacing or repairing the wing spar attachment fittings before reattaching the wings. Although it entailed a massive amount of work, the latter method negated the need for repeated inspections.

At this time, in 2003, *The Pink Lady* was based at Orly, near Paris, in a large and well-equipped maintenance hangar near to the Air France maintenance facility. After the merger of Air France and Air Inter, the team received a lot of help from Air France in the form of equipment and tools that were declared unfit for commercial use, but were perfectly serviceable for the team. With the equipment and assistance close by, the decision was therefore taken to separate the wings from the fuselage and remove the wing spar terminals for inspection. To facilitate this a large crane had to hold the wing, including the two engines. The wing itself was supported by a special hoisting sling which took the weight evenly and held it level. At the same time, the fuselage needed to be supported by jacks. The slings had to be specially made from Boeing drawings by a shipyard engineering company. It was probably one of the few occasions when a B-17 had its wings removed since the war. New wing spar terminals were manufactured, again to Boeing specifications, and the wings and fuselage were reassembled. The work took five months to complete. It was a large and costly undertaking but resulted in *The Pink Lady* becoming as strong again as the day she was built and structurally, at least, she has many more years flying ahead.

The future

The team have never received any form of official government support, probably due to the fact that the B-17 was not a type operated by the French military. Although the aircraft is in fine mechanical condition, unfortunately the economic prospects do not augur as well for the continued operation of *The Pink Lady*. The basic consumables amount to approximately 160gal/hr of aviation fuel. In recent years, the rising cost of fuel and a large increase in insurance premiums (due to the aircraft being re-categorised as an airliner for insurance purposes) has led to running costs spiralling. Thankfully, spare parts are not such a problem for the team, because when the aircraft was purchased from the IGN it came with an enormous amount of spares such as engines, turbos and tyres.

Appearances at air shows helped to pay for some of these costs, but air show bookings can never recoup the cost of maintaining the aircraft. Air shows themselves are becoming more complex and costly to organise, so consequently there are fewer being arranged, and so *The Pink Lady*, although often the star of the show, has been chasing a dwindling market.

In 2006 the hangar at Orly, frequented for many years by FTV and *The Pink Lady,* was due to be demolished. This, coupled with difficulties gaining access at Orly owing to more stringent security requirements, forced FTV to seek a new

BELOW A proud moment for the crew when they meet American Air Force legends Chuck Yeager and Bud Anderson. *(Michel Bézy)*

The six missions of *The Pink Lady*

RIGHT On 8 April 1945, 846 was among the Forts that bombed railway marshalling yards in northern Germany. Also on the mission was the 91st BG's 128-mission veteran B-17G, *Wee Willie*. The 322nd BS Fort took a direct hit from flak, tearing off its left wing and causing it to plunge out of control before exploding. *(USAF)*

BELOW A box formation of Fortresses from the 398th BG heads for its target, the marshalling yards at Neumünster, on 13 April 1945. *(USAF)*

uring the dying weeks of the Second World War, 44-8846 flew a total of six combat missions over Europe. Flying from its base at Polebrook, Northamptonshire, the aircraft was operating at a time when the Allied air forces had achieved near domination in the skies over Europe. It was certainly less hazardous than a year or so before, when USAAF bomber crews were unlikely to complete a 30-mission tour. Even so, flak remained an ever-present danger and there were new forms of enemy resistance to the American bomber formations. The potent Messerschmitt Me 262 jet fighter wrought havoc in among the bomber streams, and Luftwaffe *Rammkommando* pilots flew their only – and last – mission on 7 April 1945 to bring down American bombers by ramming their Me109s into them.

On 8 April 1945, 846 set off to bomb the marshalling yards at Halberstadt in central northern Germany. The aircraft was one of a force of 218 that attacked the town in advance of the invading ground forces. For this mission the 351st BG fielded 38 aircraft and 37 returned. The following day the target was the Luftwaffe airfield at Fürstenfeldbrück, in southern Germany. On this occasion, two aircraft were lost from the 38 despatched.

On 14 April the 351st attacked pockets of enemy troops at Royan in south-west France, who were preventing the Allies' use of the port of Bordeaux. No losses were experienced on this mission. The 16 April brought another deep penetration mission, this time to Regensburg, where the 351st was part of a force of 74 bombers that attacked a railway bridge. On this day the group suffered its final aircraft loss of the war. Two days later 846 flew to Traunstein, near Salzburg, where USAAF bombers attacked the railway marshalling yard and electrical transformers.

The final mission of the war for 846 and for the 351st was to Brandenburg near Berlin on 20 April when the group attacked marshalling yards. Thus ended the combat career of 44-8846 – a brief but historic spell for one of the very few surviving Fortresses to fly actual combat missions.

home for their aircraft. A temporary base was found at the airfield of St-Yan, before moving again in October 2007 to the old Dassault test airfield of Melun-Villaroche to the south of Paris. Here, the aircraft was carefully maintained, safe in the capable hands of the FTV engineers, who between them have many years of experience working in the aviation industry.

The story does not end here, though. Jean Salis, the last founding member of the FTV and the owner of La Ferté-Alais collection near Paris, always had plans to move the B-17 to a purpose-built hangar at the Museum of La Ferté-Alais.

In March 2010 *The Pink Lady* made what might have been her final flight, to the airfield at Cerny, home of La Ferté-Alais. Back in 1985, when the aircraft was purchased from the IGN, it was intended only to fly her for five years before displaying her at La Ferté-Alais. Due to the success of the FTV in operating the aircraft, this plan was deferred for a further 20 years. It is now planned for the aircraft to continue to be maintained in an airworthy condition at Cerny and undergo a three-year maintenance programme to attend to work that needs attention. The aircraft will be on public display during this period, something that has not been possible before. It is intended that *The Pink Lady* will one day fly again to commemorate the sacrifices of the Second World War. As the last surviving combat veteran Fortress of that conflict, who could disagree?

Whatever the future holds, *The Pink Lady* will continue to inspire future generations and to commemorate the struggles of the past.

ABOVE The two pilots who displayed 846 for many years – André Domine (left) and Michel Bézy in front of (the then named) *Lucky Lady*. *(Michel Bézy)*

Chapter One

The B-17 Story

The B-17 may never have gone into production, so fraught was its development at a time when the US Navy and US Army Air Corps were battling one another for funding, and when many in the US wanted to keep out of any future war. The Boeing Airplane Company financed development of a prototype to an AAC specification, which called for a 'multi-engine' aircraft. Boeing realised that to beat the competition their design had to incorporate the radical feature of four engines.

OPPOSITE The Collings Foundation's airworthy B-17G, 44-83575, *Nine O Nine*, is named in honour of a 91st BG, 323rd BS Fort of the same name, which completed 140 missions. *(Patrick Bunce via Keith Wilson)*

The development of the B-17

Up until 1934 the Boeing Airplane Company had built reliable, all-metal monoplane designs such as the Model 247 airliner that had a reputation for solid construction. Nothing with more than two engines had been ventured. In early 1934 the company, keen to secure government backing in hard-pressed economic times, won a development contract for an experimental long-range bomber, known as the XB-15 to the AAC, but to Boeing it was the Model 294. The new aircraft was certainly radical. With a wingspan of almost 150ft, it would become the largest and heaviest aircraft built in the US to date. It would, of course, require four engines.

In the event only one prototype Model 294 was ever built, but its design configuration was clearly influential in the aircraft which preceded it into the air – the Model 299, which would eventually become the B-17.

Model 299

Approval for the Model 299 was given by the Boeing board on 26 September 1934 and a design team was appointed, led by Boeing President Clairmont Egtvedt with Giff Emery as Project Engineer and Edward C. Wells who was only 24 at the time, as Assistant Project Engineer. However, it is Ed Wells' name alone that has become associated over the years with the B-17 programme.

The AAC's 1934 specification, for which the Model 299 was developed, required a top speed of 250mph at 10,000ft, a cruising speed of 220mph, with an endurance of 10 hours and an engine-out altitude of 7,000ft while carrying what was termed 'a useful load'.

When the new aircraft was unveiled to the press in July 1935, it looked like nothing that had been seen before. It resembled, in the style of the times, a beautiful Art Deco object with its highly polished aluminium skin, sleek and tapered fuselage and graceful curved wings. So impressed was one reporter from the *Seattle Times* that he referred to the aircraft as a 'fifteen-ton flying fortress'. Boeing liked the name and registered it as a company trademark.

The Model 299 was powered by Pratt & Whitney Hornet radial engines housed in aerodynamically streamlined nacelles developing 750hp each; these were a proven and reliable power plant. They drove three-bladed, constant-speed propellers.

The structure of the cantilever wing was adapted from the XB-15 and the Model 247 designs, consisting of two main spars built up using a truss design. The spar trusses were aluminium square section tubes placed vertically and diagonally and riveted to web plates to form single units. Connecting the spars, metal ribs running chord-wise were also stiffened using the truss design, in this case the tubes were of a round section. Over this primary structure a layer of corrugated aluminium stiffening was riveted, running span-wise, before the outer aluminium skin was attached over this. The complete stressed skin structure was immensely strong.

The fuselage, constructed as a semi-monocoque structure with a circular cross-section, had a maximum circumference in the cockpit and bomb-bay, gradually tapering down to a point at the tail. A built-up decking structure extended from the cockpit to halfway along the fuselage.

The fuselage consisted of bulkheads and circumferential stiffeners or frames connected by longitudinal stiffeners and load-bearing longerons. Over this was riveted the aluminium skin. In the areas where the loads were greatest, the fuselage frames were placed closer together, such as in the centre section containing the bomb-bay.

The empennage (or tail plane) consisted of all-metal fin and stabiliser, with fabric-covered rudder and elevators. The fin and rudder assembly above the fuselage came to be known as the 'shark fin', in order to distinguish it from later versions that underwent a major redesign of this area.

Defensive positions were included in the fuselage in the form of teardrop-shaped gun blisters, one on each side of the rear fuselage, one below and another in the rear of the upper decking structure. A further gun position was installed within the braced nosecone, consisting of a gun cupola which could be rotated through 180°. All guns were manually operated and could be either .30 or .50-calibre.

The side and top-fuselage gun blisters were eliminated and the side blisters were replaced by flush-mounted windows, which could be opened to allow the weapons to be fired. Extra gun-mounting sockets were added to the nosecone and the side windows of the nose compartment, and some armour plate was installed around crew stations in an effort to make the aircraft more combat-worthy. An improved version of the Wright Cyclone engine, the R1820-65, which developed a maximum 1,200bhp at take-off, was installed in the B-17C. A total of 38 B-17C aircraft were delivered between July and November 1940.

From this number, 20 airframes were diverted from the production line for the Royal Air Force (RAF), who planned to use the aircraft on daylight bombing missions. Known to the RAF as the Fortress I, the aircraft differed in having minor changes to equipment and armament, as well as wearing British markings and serials. But the most important modification, subsequently incorporated on all future versions of the B-17, was the installation of self-sealing fuel tanks. These

were constructed in two parts – an inner lining that contained the fuel, encased in an outer layer of natural rubber. If a fuel tank was punctured by a bullet or shrapnel, the leaking fuel would start a chemical reaction when it came into contact with the outer layer, causing the natural rubber to swell and seal the hole. This proved to be a vital modification that made the aircraft much less likely to explode when receiving direct hits in the wings during combat.

ABOVE One of 20 B-17C aircraft acquired by the RAF as the Fortress I, AN530 was operated by 90 and 220 Squadrons. It had the distinction of being the first Fortress to engage with enemy aircraft on 2 August 1941. *(IWM HU74972)*

LEFT This Fortress I carries the incorrect RAF serial, AM528, applied prior to its delivery to the UK. As AN528 its service career was short, lasting barely a month before it burnt out as a result of an accident at Polebrook in July 1941. *(via Peter Brown)*

B-17D

Further improvements to the design led to the B-17D series. This was very similar to the B-17C series, the only outward difference being the hinged cowl flaps behind the engine cowlings. Internally, improvements were made to the oxygen system and an up-rated 24-volt electrical system was installed. Armament was increased by pairing the .50-calibre weapons in the belly 'bathtub' and the radio compartment. Delivery of 42 B-17D series aircraft took place between April and September 1941. Eventually, the remaining B-17Cs still on the US inventory were subsequently modified to B-17D specification, but excluding the addition of the cowl flaps, and from April 1941 B-17s began to be painted in camouflage colours. Most B-17Ds went to bomb groups based in the Philippines and Hawaii and as a result many were caught up in the Japanese attack on Pearl Harbor, which led to the destruction of a large number of the bombers on the ground.

B-17E

The aircraft had by this stage reached a point where the basic design could not be adapted any further to improve its defensive capabilities. What was required was a radical reworking of the airframe. To its credit Boeing had recognised this before the unfortunate experience of the RAF with the Fortress I (*see Chapter Two*) and had begun a redesign of the airframe, which was to totally change its appearance and significantly improve its defensive capability. The rear fuselage aft of the wing root was greatly enlarged and extended to provide room for a gun position in the extreme tail. This had been a vulnerable area in the earlier versions and the provision of a gunner with two .50-calibre weapons did much to redress this weak spot. Additionally, the horizontal tail surfaces were enlarged, but most significantly the vertical fin and rudder were totally redesigned and enlarged, the fin blending into the fuselage in a long taper as far forward as the wing root. Below the fuselage a ventral remotely operated Sperry turret was fitted, housing two more machine guns, the gunner sighting his guns through a periscope just aft of the turret. Immediately behind the cockpit another turret was installed. This Sperry-designed unit was

manned by the flight engineer and was also equipped with two .50-calibre weapons. These changes became necessary not only to improve the defensive firepower of the aircraft (the earlier aircraft were Fortresses in name only, being conspicuously vulnerable to fighter attack), but also to improve the stability of the B-17E, especially on bomb-release at high altitude.

The B-17E was virtually a new aircraft. The link with the art deco style and 'shark fin' of the Thirties design had been broken and the redesign gave it a much more purposeful, fuller appearance. The huge dorsal fin and rudder at first appeared ungainly to Forties eyes, but quickly was seen to bestow a graceful curved elegance to the airframe. Today the profile of the B-17 is regarded as the classic shape symbolising American airpower in the Second World War. The redesigned tail had been developed from another Boeing aircraft, the Model 307 Stratoliner, a design that had suffered from directional stability problems which an enlarged dorsal fin cured. The Model 307's fin design, adopted for the B-17E, transformed the aircraft into a superbly stable bombing platform. Many hundreds of smaller design changes made the 'E' the first truly combat-ready Fortress.

First flight of the B-17E took place in September 1941 and production ended in May 1942, with a total of 512 examples built. The only major change in the production run occurred after the 112th example with the replacement of the remotely sighted ventral turret by the Sperry ball turret. This unit was designed to house the gunner, machine guns and gun sight in a self-contained unit that gave 360° covering fire to the underside of the aircraft. Although conditions for the gunner were cramped, it proved a far more practical defensive position than the earlier periscope-sighted unit.

B-17F

In May 1942 the B-17F series entered production, incorporating around 400 internal changes and improvements from the E. Outwardly the only distinguishing feature was a longer, frameless nosecone. Significant modifications included a change to wider 'paddle-blade' propellers, which improved high-altitude performance. Additional range was provided by the fitting of nine extra fuel

The data block

Every B-17 carried a data block, stencilled in one-inch-high letters just forward of the cockpit on the left-hand side. The information in this block gave details about the individual aircraft concerned.

The first line denotes the procuring service, ie, US Army Air Force (after June 1941, but prior to that the US Army Air Corps).

Next is the aircraft type, model and series designation, in this case B-17G. Here, the 'B' stands for 'Bombardment' type; '17' is the model number of this type contracted for by the USAAF; and 'G' is the suffix indicating the series designation of the original model. Series designations were added when there were a number of changes and improvements to the model. Although commonly referred to as 'B-17G model' for example, the correct term is 'B-17G series'.

Following this is the block number, which gives details of the modification state of the aircraft and the manufacturer. The '-35-BO' refers to this aircraft being part of production block 35, built by Boeing in Seattle. Due to the large number of changes and improvements that took place during manufacturing and the fact that there were three companies building the B-17, production block numbers were brought in to keep a record of these changes. The two other manufacturers who built the B-17

under licence were Douglas Aircraft at Long Beach, California, using the designation 'DL', and Vega Aircraft, a subsidiary of Lockheed in Burbank, California, with the designation 'VE'. Collectively known as the BVD Pool, even though, or perhaps in spite of, to every American this was the name of a famous brand of underwear!

The serial number is the military number allocated by the acquiring service; it is a different number from the manufacturer's own constructor's number. The serial in this example, '42-31983', refers to the Fiscal year (or financial year or budget year) in which the aircraft was contracted, in this case the fiscal year running from 1 July 1941 to 30 June 1942, and this was the 31,983rd aircraft contracted for. The serial was allocated when the aircraft was ordered. Delivery could be much later; in this particular case it did not take place until January 1944. The serial was also applied to both sides of the vertical fin in a condensed form where the first digit and the dash are omitted, ie, '231983'.

The last line of the data block is always the same and nearly always incorrect. The crew weight of 1,200lb only applied to the early B-17s with a six-man crew (200lb per man). From the B-17E onwards, the typical crew numbered ten; however, the data block was never changed to reflect this.

BELOW The data block for *Mary Alice*. (Graeme Douglas)

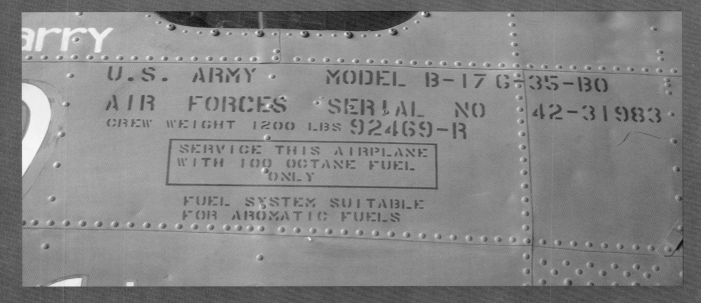

ABOVE A B-17F pictured over the Washington area, probably in early 1943. It is unusual in having external racks fitted, which allowed the aircraft to lift a greater weight of bombs, but at the expense of speed and range. *(IWM NY602)*

RIGHT A Lockheed-Vega-built B-17F runs up its four Wright Cyclone engines before delivery to a combat unit. *(Lockheed)*

cells in each wing, between the outer wing ribs, popularly known as 'Tokyo' tanks. These held another 1,080gal of fuel. Combat experience soon showed up deficiencies in the aircraft's lack of forward firepower and a tail-heavy balance condition. To improve the forward defences, modifications were made to enable extra guns to be carried in the nosecone. Additionally, the side windows were adapted to mount a machine gun on each side of the nose, these later being modified into bulged 'cheek' gun windows.

The B-17F was the first version to be built by what became known as the BVD Pool, with production shared between the aircraft manufacturers Boeing, Vega and Douglas. This was a major step towards increasing production output, which was to prove vital in maintaining the flow of replacement aircraft to the European theatre when losses were at their greatest. Each of the three manufacturers introduced modifications on the production lines and all similarly modified aircraft were allocated to the same manufacturer's production block number.

A total of 3,405 B-17F aircraft were constructed by the three manufacturers between May 1942 and September 1943.

ABOVE Workers at the Lockheed-Vega factory watch the roll-out of the first Vega-built B-17 in May 1942. *(Lockheed)*

"KEEP 'EM FLYING"
IS OUR BATTLE CRY!
...
DO YOUR PART
FOR
DUTY - HONOR - COUNTRY

LEFT The B-17 Flying Fortress features in this recruitment poster produced by the US War Department in 1941, appealing to the patriotism of every American citizen. After the Japanese attack on Pearl Harbor on 7 December, the War Department asked US industries, trade organisations, and advertising agencies to promote the slogan 'Keep 'Em Flying' as a patriotic service. A Hollywood movie called *Keep 'Em Flying*, starring the comedy team of Abbott and Costello, was released on 28 November 1941 to coincide with the 'Keep 'Em Flying' Week.

Boeing B-17G Flying Fortress – technical specification

(Variations for *The Pink Lady* are indicated in brackets)

Dimensions and weights

Wingspan	103ft 9.38in
Length	74ft 8.9in with tail stinger, 74ft 3.9in with shorter nosecone
Height	19ft 1in (tail down)
Horizontal stabiliser span	43ft 0in
Wheel track	21ft 1.52in
Wing area	1,277.5sq ft (net, with flaps and ailerons)
Maximum take-off weight	65,500lb (PL – 46,300lb)
Maximum fuel load	3,600 gallons if bomb-bay tanks fitted (PL – 2,780 gallons)
Oil tanks	36.9 US gallons each with 10 per cent expansion space

Speed limitations

Maximum speed	305mph (PL – 225mph)
Max flap-lowering speed	147mph

Centre of Gravity limitations

The centre of gravity must fall between the following limits:
Forward: 20 per cent MAC, which equates to 270.8in aft of the datum.
Aft: 32 per cent MAC equating to 292.1in aft of the datum.

Engine limitations

	Rpm	Manifold pressure (in Hg)	Cyl head °C	Oil °C
Maximum take-off 5 minutes max	2,500	47.5	260	88
Maximum cont climb	2,300	41.5	232	88
War Emergency Power 1,380hp 5 minutes max*	2,500	54.0	260	95

*Only available when water injection systems were installed on the engines.

(For engine limitations of *The Pink Lady* see Chapter Seven)

B-17G

The final production series was the B-17G, externally identified by a remotely controlled, electrically powered 'chin' turret beneath the nosecone. Somewhat confusingly, the last few B-17Fs built by Vega and Douglas were fitted with the chin turret before the production designation changed. All B-17s with chin turrets were designated as 'G' by the USAAF. Bendix built the all-electric-powered chin turret, originally developed for the B-25 medium bomber. It was first trialled on a heavily armed escort version of the B-17, the YB-40, which proved unsuccessful. The chin turret concept, however, was introduced into B-17 production.

The early production G's were completed without the cheek gun installations, in the belief that these would no longer be necessary now that the power turret defended the front of the aircraft. Combat units found otherwise, and these were reinstated on the production lines. This took the maximum number of guns carried by the B-17 up to 13 – the aircraft could finally live up to the name of Flying Fortress.

Other important changes introduced into the G series production were the staggering of the waist windows to give gunners more room in which to operate. The windows were also glazed over to reduce the icy blast of air from affecting the gunners in the rear of the aircraft.

A major improvement to engine handling came about with a development for the turbo supercharger controls. Earlier models controlled the speed of the turbos, and consequently the engine boost, by a mechanical linkage in the cockpit. This linkage was connected to hydraulic actuators operating the turbos. In the sub-zero temperatures experienced at high altitude, the oil in the hydraulic system tended to thicken, leading to sluggish operation of the turbos. Pilots had to regularly exercise the controls to keep them functioning properly, although instances of loss of control of the engine boost with this system did occur. With the introduction of electrical speed control, this rather clumsy system was eliminated and replaced by a single knob for the pilots to control the turbo boost. Each individual unit

RIGHT Two USAAF officers in front of an early B-17G, identified by its twin-gun chin turret. The guns are fitted with flash suppressors to prevent the blast from cracking the plastic nosecone. Note the lack of cheek guns, thought to be unnecessary with the introduction of the chin turret. They were soon reintroduced. *(ww2 images)*

was regulated by an electrical regulator and waste gate actuator, a modification that was praised by pilots and considered to be one of the most important improvements in the production run.

Also during G series production, the tail-gun position was redesigned to provide more room and better visibility for the gunner, along with more ammunition and improved sighting for his guns. United Air Lines developed the new installation at their modification centre at Cheyenne, Wyoming, and it became universally known as the 'Cheyenne' tail. It was supplied as a retrofit unit to overseas maintenance depots and many aircraft had the Cheyenne fitted in the field before the change was incorporated on production lines. First delivery of the G series was in September 1943 and Vega delivered the last examples in July 1945.

At the peak of production in April 1944, Boeing, Vega and Douglas were completing 16 B-17Gs a day. Between the three plants they produced a total of 8,680 G series aircraft of which 4,750 were lost in combat. A total of 12,731 B-17s of all marks were produced.

As well as the production series aircraft listed above, a small number of B-17 variants were also developed. The major types are listed below.

YB-40

Designed as heavily armed escort bombers, the YB-40s had increased armament, ammunition and protective armour but carried no bombs. The idea was that these aircraft

RIGHT A Lockheed-Vega employee signs her name on the fuselage of B-17G, 44-8628, named *March of Dimes*. *(Lockheed)*

RIGHT The awesome scale of the American war machine can be judged by these views of the Boeing plant in December 1942, when B-17F production was in full swing. Bomb-bays are being constructed in their jigs. The platform enabled two sets of workers, one above the other, to work on the assemblies at the same time. In the background, rear fuselage sections are being fitted out. *(Library of Congress D-008449)*

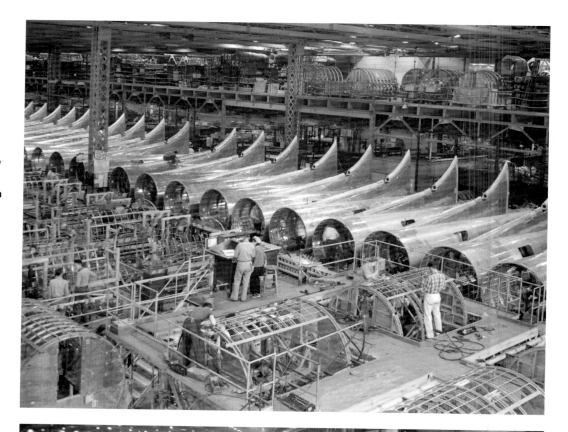

RIGHT Forward fuselages are constructed from the sub-assemblies of the nose, cockpit, bomb-bay and radio room. Aluminium skin is then attached to the structure. The fuselage on the left has had the top decking structure added to the circular fuselage. *(Library of Congress D-008211)*

LEFT A Fortress II of 206 Squadron, RAF Coastal Command, takes off from Lagens on the Azores for an anti-submarine patrol over the Atlantic. Three Coastal Command squadrons operated the Fortress: 59 Squadron (Fortress II, December 1942 to April 1943), 206 Squadron (Fortress II, July 1942 to March 1944), and 220 Squadron (Fortress I, II and III) from December 1941 until April 1945. *(IWM CA87)*

but could be blamed – at least partly – on the fact that the British Fortresses carried a Sperry bomb sight, a less sophisticated design than the Norden developed by the US Navy, and still a closely guarded secret. However, the lessons learned from these disappointing experiences were used to improve the later versions of the B-17 in USAAF and RAF service.

No 90 Squadron continued to operate its Fortresses over Europe for two more months, but with little success. Later it had a detachment operating in the Middle East but in February 1942 the squadron finally gave up its Fortresses for Short Stirlings.

It was a more successful story for the RAF when three Coastal Command squadrons operated the Fortress I, II and III on anti-

submarine patrols over the Atlantic from bases in the UK and the Azores. At the war's end the Fort was in use by two RAF meteorological squadrons, one air-sea rescue squadron and an anti-submarine squadron.

A small number of B-17s were also used by two squadrons in the RAF's 100 (Bomber Support) Group in the specialised role of electronic countermeasures. Their purpose was to try and lessen the heavy losses suffered by Bomber Command in the strategic air offensive by jamming the Luftwaffe's air defence systems during bombing missions. RAF Fortresses were fitted with arrays of powerful radio transmitters and receivers to jam enemy radio, radar and signals, and for electronic intelligence gathering.

BELOW, LEFT AND RIGHT 250lb depth charges are loaded into a 220 Squadron Fortress IIA at Benbecula in the Outer Hebrides during May 1943. *(IWM CH11101 and IWM CH11102)*

BELOW The B-17 was able to withstand enormous battle damage and still return to base. Here, crew chief, M/Sgt Dewey Christopher stands in front of a hole ripped open in the fuselage by flak after a raid on Seelze near Hanover on 14 March 1945. The Fortress, 43-38852, EP-N, of the 100th BG, was brought back to Thorpe Abbots, Norfolk, by Lt Ed Aubuchon. Waist gunner, Cpl Garland Miller, was killed. *(IWM 52946)*

planned at the same time to draw fighters away from the main bomber forces.

The plan began to unravel when many of the East Anglian airfields were fog-bound on the day of the mission. Although the Regensburg force (4th Bombardment Wing, made up of seven B-17 bomb groups) departed on time, the 1st Bombardment Wing (made up of nine B-17 bomb groups) destined for Schweinfurt was delayed for 3½ hours until the fog lifted sufficiently. This was to prove critical, as the German fighters, after attacking the first force, had time to land, re-fuel, re-arm and then take off again to meet the delayed bombers. The result was carnage as the bomber stream was attacked relentlessly as it pressed deeper and deeper into enemy airspace beyond the range of the fighter escorts. After reaching and bombing the target the Regensburg force headed onwards to bases in North Africa. The

force attacking Schweinfurt, however, had to turn and head back the way they had come and suffered the same onslaught on their return journey. In total that day, out of 376 B-17s despatched, 60 were lost and many more were badly damaged. As well as the record number of US aircraft losses that day, the American gunners claimed the huge total of 288 enemy aircraft shot down, and although grossly exaggerated, the figure indicates the degree of intensity with which this bloody air battle was fought. The true total was 27 Luftwaffe aircraft lost on the day.

The results of the attack on the Messerschmitt factory at Regensburg were assessed as being highly successful as the bombers found their target in near perfect visibility. The raid on Schweinfurt was not such a success. Although major damage was inflicted to the ball bearing plants, production was affected only for a short time. However, it did lead the Nazis to reorganise their vital war industries, dispersing manufacturing to numerous smaller, better-hidden factories rather than concentrating them in large and vulnerable complexes.

Many of the tactics developed for the Eighth AF were pioneered by Colonel Curtis LeMay, the first commander of the 305th Bomb Group (BG). He advocated the use of 'lead bombers' whose crews underwent intensive training to develop their skills in accurate bomb dropping. These crews were to lead the formation and on dropping their ordnance, all the following aircraft would do the same. The theory was that more accurate and concentrated bombing could be achieved by following the signal from the most experienced and highly trained crews, than allowing each bombardier to drop individually.

Sometimes a mission would work exactly to plan. The raid on the Focke Wulf plant at Marienburg on 9 October 1943 was described as 'the bombing of the year' when 96 B-17s achieved highly accurate results, causing severe damage to the plant. This was achieved by using a lead aircraft and bombardier, in almost perfect conditions with excellent visibility and little flak, allowing bombing to take place at a much lower altitude than normal, between 11,000 and 13,000ft.

But this was the exception to the rule.

More often the bombers faced onslaughts from enemy fighters as well as highly accurate flak. Added to this were the constant delays, diversions and aborts to missions caused by the changeable weather over north-west Europe. Meteorological conditions were often marginal for operations and on many occasions the bombers reached the target only to find it obscured by cloud, necessitating a secondary target to be chosen, or for a target of opportunity to be found. The example of the Marienburg raid also illustrates that more accurate bombing was achieved from lower altitudes. Most missions were flown at around 25,000ft and accuracy was traded against the need to avoid the worst of the flak. The ideal of pinpoint precision bombing by each aircraft that had been the goal of the strategic bombing enthusiasts in the early stages of the war had been sacrificed to expediency.

ABOVE **A huge pall of smoke hangs over Marienburg as B-17s of the 94th BG begin the long flight back to base on 9 October 1943.** *(USAF)*

Norden bomb sight

ABOVE **The Norden bomb sight installed in** *The Pink Lady*. **The cylindrical unit at the top right is an optional reflex sight, which provides the bombardier with a greater field of view for sighting.** *(Graeme Douglas)*

The B-17 Flying Fortress and B-24 Liberator bombers of the USAAF were both equipped with the Norden bomb sight. Three types were used – the M-7, M-8 or M9, all being similar in design. It was thought that armed with such an accurate bomb-aiming device, pin-point, strategic bombing could bring a swift end to the war.

The Norden was highly advanced for its time. With a complex mechanical gyroscope system it was able to compute the correct release point of the bombs based upon information fed to it by the bombardier. Using parameters such as bomb ballistics, aircraft altitude, speed, and drift, the sight was theoretically able to release the bombs automatically and be accurate enough to place them within a 100ft radius from an altitude of 20,000ft. In battle conditions, though, this sort of accuracy was rarely achieved. During the latter part of the war, bombing tactics developed so that only the lead aircraft in a formation would use the Norden over the target, the following aircraft all bombing visually on the release of the leader's bombs.

The sight was constructed in two parts – the lower section contained the stabilising gyroscope, which was linked to the C-1 autopilot system and allowed the bombardier to control the lateral movement of the aircraft through the bomb sight during the bomb-run. The upper part was the sight proper, which comprised the optical head and its own stabilising system. During the war, efforts were made to keep the details of the sight secret and it was either removed from the aircraft or covered up between missions and is rarely to be seen in period photographs.

The B-17's bomb-bay

ABOVE Armourers load 500lb bombs into a B-17's bomb-bay. The bombs are winched by another armourer above and attached to one of four racks. *(ww2 images)*

BELOW An under-wing bomb carrier is loaded with a 1,000lb bomb. *(Jonathan Falconer collection)*

The bomb-bay of the B-17 occupies the space in the fuselage between the two wing spars. This meant it was not possible to enlarge it and the size and shape of the bomb-bay remained the same throughout the aircraft's production life.

When the Model 299 flew in 1935, it did so as a long-range maritime reconnaissance bomber carrying a bomb load that (for the time) appeared sufficient. World events moved on, but the Fortress bomb load could not be increased to face the threat from Nazi Germany.

Because of size restrictions, the largest bombs that could be accommodated weighed 2,000lb. Bombs were winched up by hand and suspended horizontally on four racks – two racks on either side of a crew catwalk that passed between them. Smaller bombs could be suspended above the larger ones that were carried at the bottom of the bay, but the average load for a mission in Europe was around 4,000 to 5,000lb. Compared to other allied aircraft this seemed a lot of aircraft for a comparatively small bomb load. The B-24 Liberator, with its two bomb-bays, could average about 1,000lb more than the Fortress. The British heavies, the Stirling, Halifax and most notably the Lancaster, with their bomb-bays extending for half the length of the fuselage, regularly carried loads of 13,000 and 14,000lb. Extra weight could be carried by the B-17 using under-wing racks; however, this was at the expense of speed and fuel consumption and it was not common practice.

The release mechanism for the bombs sometimes malfunctioned, leaving lone bombs still hanging on the racks after the rest of the load had been dropped. In these cases it took some courage from a crew member to negotiate the catwalk with the bomb doors still open, and then try to kick or prise the bomb from the rack.

On 14 October 1943, the Eighth again visited Schweinfurt. In total 291 B-17s returned to the bearing plants deep in the heartland of the Reich. Things were no easier for the bombers this time despite a diversionary raid by a small force of B-24 Liberators. Three of the factories were badly hit but a total of 60 aircraft were lost on this raid, 17 more crashed on their return or were written off due to battle damage, and more than 120 were damaged. Intelligence confirmed that huge amounts of damage had been done to the bearing factories; output fell dramatically and did not return to normal for six months. Again, in the confusion of battle, huge claims were made for the number of Luftwaffe aircraft shot down by the B-17 gunners. The total claims for the previous five raids amounted to over 700 enemy aircraft destroyed, with more than 300 claimed as probables. These claims, had they been correct, would have accounted for the entire Luftwaffe fighter force. Although the figures were widely believed at the time, it soon became evident to those fighting the air battles that the German Air Force was far from defeated. It was also very clear that mass formations of bombers could not defend themselves from fighter attacks. What were needed were long-range escort fighters to protect the bombers all the way to and from the target.

Up until this time the American fighters, the P-47 Thunderbolt and the P-38 Lightning, were only able to escort bombers for part of their deep penetration missions. Their lack of range left the bombers very exposed once the fighters turned for home. However, on 16 December 1943 the first fighter group equipped with the long-range P-51B Mustang (the 354th FG) began flying escort duty for the B-17s. The P-51B was fitted with an American Packard-built version of the British Rolls-Royce Merlin engine. The combination of the aerodynamically efficient airframe and the excellent engine produced an aircraft of fine performance with approximately half the fuel consumption of the P-38 and P-47. When fitted with external drop tanks the P-51's range increased further, giving it the ability to fly up to

BELOW B-17Gs from the 401st BG, based at Deenethorpe, Northamptonshire, drop 500lb bombs. Note the smoke markers indicating the target. *(Author's collection)*

ABOVE B-17Gs from the 381st BG are escorted by a P-51B Mustang on a training flight over England in the summer of 1944. When the P-51 was re-engined with the Rolls-Royce Merlin engine and fitted with long-range tanks, it was capable of escorting Forts and Liberators all the way to Berlin and back again. *(USAF)*

BELOW The effects of a direct hit from flak. B-17G, 43-38172, *Lovely Julie*, of the 398th BG, 601st BS, received a direct hit from an 88mm flak shell over Cologne on 15 October 1944. She was nursed back to base at Nuthampstead, Hertfordshire, by her pilot, 1st Lt Lawrence M. DeLancey. All survived except the bombardier who was killed instantly. *(USAF)*

1,500 miles from base to target and back. With the Mustang build-up during 1944 the heavy bombers at last began to get fighter protection all the way to the target and home again, even on the deepest missions into enemy territory.

One target previously not attempted by the B-17s and B-24s of the Eighth was the capital of the Third Reich itself, Berlin. Although the RAF had attacked Berlin under the cover of darkness many times since 1940, the USAAF still considered it too distant and too well defended to attack in daylight, until the advent of the P-51 Mustang fighter made it possible. The first American bombs fell on Berlin on 4 March 1944, dropped by a small force of B-17s in a raid which had actually been recalled due to bad weather. The bombers continued to the target as they had not received the recall signal. Fighter opposition on this occasion was comparatively light. However, the first full-scale raid took place two days later on 6 March.

During the winter months of 1943–44, the B-17 squadrons of the Eighth had seen much of their efforts directed against Nazi V-weapon sites in northern France, but with the build-up to D-Day their attention was also focused on disrupting the enemy's transport infrastructure in northern France and Belgium. In May, B-17s flew their first missions against oil plants in Germany and western Poland, a target system that was relentlessly attacked with success by B-17s and B-24s until March the following year.

Codename 'Aphrodite'

In 1944 the USAAF created a secret programme codenamed 'Aphrodite' as an experimental method of destroying V-weapon sites and other 'hard' targets using unmanned radio-controlled B-17 and PB4Y bombers that had outlived their operational usefulness.

At least 25 B-17s were stripped of their combat armament and other non-essentials, fitted with radio controls and packed with 20,000lb of high-explosive Torpex. 'Aphrodite' B-17s were flown manually by two pilots who bailed out after take-off and the drone was 'flown' to its target using radio controls by a 'mother ship', usually another B-17 or B-24. 'Aphrodite' B-17s flew thirteen one-way missions to France and Germany between August 1944 and January 1945, but not one aircraft hit its target. The project was abandoned.

ABOVE The nose hatch of the drone B-17 was enlarged and fitted with a wind deflector to make it easier for the pilots to bail out. The tank below the fuselage contained a chemical which emitted smoke to enable the controller aircraft to see the unmanned drone. *(Ian McLachlan)*

LEFT AND BELOW *Gremlin Gus II*, a veteran of the 388th BG, was earmarked for the 'Aphrodite' missions but she was never used. Later she was prepared for a radio-controlled strike on the German battleship *Tirpitz*. In order to load the air-dropped torpedoes necessary for the attack, the top decking was removed and the fuselage faired over. The *Tirpitz* strike never happened and *Gremlin Gus II* ended her days in a training role as the only open-cockpit B-17. *(Ian McLachlan)*

Forts over the 'Big City'

When 730 B-17s and B-24s of the 1st, 2nd and 3rd Bomb Divisions raided Berlin on 6 March 1944 in the USAAF's first full-scale daylight mission over the capital, the Luftwaffe was waiting for them. Flying a direct course to Berlin, the bomber stream tailed back almost 100 miles. The route meant the aircraft would fly some 800 miles over hostile territory for a period of about six hours, thereby giving the Luftwaffe plenty of time to assemble and plan their attacks. Although more than 800 escort fighters accompanied the bombers all the way to the target and back, the sheer scale of the raiding force meant that it would be impossible to protect the bombers at all points along the route. Based on the known German tactics of focusing their attacks on the leading formations, the fighters concentrated their numbers in this area.

Due to a radar failure in the lead aircraft, the vanguard of the bomber stream became separated from the rear and drifted south of its intended course, while the rear of the stream continued on the planned route. This was a small deviation, which gradually developed into a gap of about 20 miles. Hence the leading part of the second half of the stream was very exposed, with very few escort fighters in the area. The opposing Bf109 and FW190 fighters were vectored to the front of the formation by their controllers and bomber after bomber fell to the guns of the experienced German pilots. One bomb group in particular, the 100th, suffered terrible losses that day: 15 of its aircraft failed to return, of which 10 came from a single squadron. The group already had a reputation as a 'hard-luck outfit' from previous heavy losses; now the nickname 'The Bloody Hundredth' became attached to the Thorpe Abbots-based flyers.

The vanguard of the bomber stream eventually corrected the track error and rejoined the rearguard, so that for the attack on Berlin the air armada was a continuous stream again. Once the bombers reached the outskirts of the capital they encountered heavy flak from the city's anti-aircraft batteries. Due to cloud obscuring the primary targets, bombs were released on to either the secondary targets, or targets of opportunity, which meant some aircraft spent longer over the heavily defended areas than intended. Once they had released their loads the bombers faced the long slog home, subjected again to attacks from the same fighters and flak batteries.

In total that day, the Eighth Air Force lost 69 bombers, the highest loss rate for a single mission by the Eighth. Additionally, 11 fighters were also lost. A total of 232 men were killed in action or missing, 415 were taken prisoner, with a further 61 evading capture and making it back to Britain. Twenty-nine wounded airmen were taken from damaged aircraft that landed back at their bases. German losses totalled 66 fighters destroyed or damaged beyond repair, with 36 aircrew killed and 25 wounded. Although the figures clearly show the highest losses to be on the American side, the losses for the Germans were more serious than they first appeared. Unlike the Americans, the Luftwaffe was finding it increasingly difficult to replace experienced pilots killed in action. New pilots could not be trained quickly enough to replace those lost in combat.

The success of the Berlin mission in bombing terms was not good. Two of the three primary targets, the bearing works at Erkner and the electrical works at Klein Machnow, were not hit at all as they were obscured by cloud; the Daimler Benz aero engine plant at Genshagen was only slightly damaged. Bombs were dropped on the city through gaps in the cloud cover, with some damage being done to road and rail infrastructure, but mostly they fell on civilian areas. A total of 345 civilians were killed or missing and 363 were wounded. The effect on morale, though, was something that could not be measured. For the first time Berliners had seen massed formations of American bomber aircraft over their city in daylight; they could now see that the enemy could get through by day as well as by night. The Reich capital was not invulnerable; the Luftwaffe and the flak could not stop the bombers. For the Eighth Air Force, the taboo of Berlin had finally been broken.

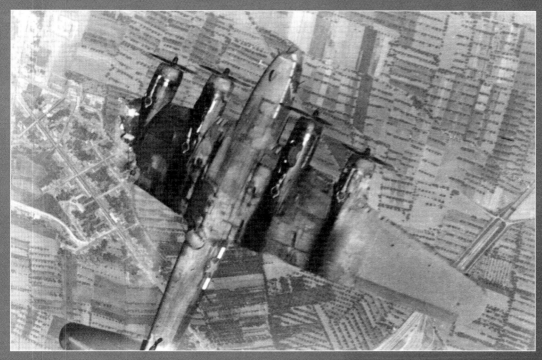

LEFT A B-17F, its starboard wing blown away by a direct hit from flak, begins its death dive. *(USAF)*

BELOW Aircraft from the 94th BG attack targets at Emden in October 1943. A normal bomb load would be 12 x 500lb bombs, although some of these sticks contain less – indicating bombs that have hung up on the bomb racks, which was a not uncommon occurrence. *(IWM EA10749)*

During the summer of 1944 the Eighth mounted a number of daylight raids on German military vehicle plants and depots, its aim being to restrict the replacement of transport lost by the Germans in Normandy. By the end of September the USAAF and RAF heavy bomber squadrons were released from regular tactical operations in support of the ground forces and resumed their round-the-clock bombing of German targets.

Bad weather in Europe over the harsh winter of 1944–45 seriously hampered bombing operations, although tactical targets linked with the ground offensive accounted for three-quarters of operations flown in January 1945. In February the strategic air assault on Germany was resumed with vigour when the German oil industry and communications targets were in the bombers' sights once again. In the closing weeks of the Second World War the devastation caused by Allied heavy bombers like the B-17 helped to seal the fate of the Third Reich. The final operations of the war flown by B-17s were humanitarian in nature, dropping food supplies to the starving population of Holland.

ABOVE The starboard cheek gun position on B-17G, 42-97385, *Shady Lady*, of the 398th BG, 601st BS. *Shady Lady* suffered engine failures returning from Ludwigshafen on 8 September 1944 and crash-landed in eastern France. *(Jonathan Falconer collection)*

These raids, combined with attacks on German aircraft factories, began to make serious inroads into fighter production and supplies of the fuel they needed.

RIGHT Bomb doors open and pulling contrails, a B-17G prepares to drop its load. *(USAF)*

B-17s over the Med and Pacific

Mainly operating from a complex of airfields in southern Italy, the US Fifteenth Air Force was formed on 30 October 1943. In the 18 months of its existence, B-17s and B-24s of the Fifteenth attacked and destroyed oil targets within their range in southern Europe and knocked out the enemy's major aircraft factories in the region.

The Fifteenth was made up from 21 bombardment groups – 6 of these were B-17 groups comprising 26 B-17 squadrons; 15 were B-24 bombardment groups with 60 B-24 squadrons.

It fought four broad campaigns: against enemy oil, the enemy air force, enemy communications, and enemy ground forces. Most vital of the Fifteenth's oil targets was the Ploesti complex of refineries in Romania, which contributed about 30 per cent of the entire Axis oil supply and an equal amount of petroleum. The Fifteenth followed up the Ploesti attacks by dropping 10,000 tons of bombs in attacks on three synthetic oil plants in Silesia and one in Poland, reducing their combined production by February 1945 to 20 per cent of what it was in June 1944.

With repeated fighter and bomber attacks, the Fifteenth crippled the enemy's transportation system across half the area of enemy occupied Europe. On occasion it helped disperse enemy counter-attacks and spearheaded the advances of the Allied armies at Salerno, Anzio, and Cassino.

By launching devastating attacks on Wiener Neustadt and Regensburg in Austria, two of the three main centres for fighter aircraft manufacture, the Fifteenth helped materially in the attainment of European air supremacy over the Luftwaffe.

As well as its commitment to the counter-air force and oil campaigns, the Fifteenth also attacked enemy communications and transportation targets far behind the front lines, and disrupted supply movements from industrial centres over an 800-mile radius from the Italian airfields.

On 15 April 1945 the Fifteenth put up a record 93 per cent of its available aircraft to soften up the approaches to Bologna in one

ABOVE B-17F, 41-24457, leaves the target after a strike on Japanese shipping off Gizo Island, Solomon Islands, in October 1942. The B-17F, named *The Aztec's Curse,* was assigned to the 5th BG, 31st BS. *(USAF)*

of the final missions of the Italian campaign.

Only five USAAF B-17 bomb groups operated in the south-west Pacific theatre during the Second World War, and all had converted to other aircraft types by mid-1943. The longer-ranged B-24 Liberator was most favoured for its higher speed and greater bomb load. In addition, heavy bomber losses in Europe were reaching such a level that the entire B-17 production was urgently needed for replacements and training in that theatre.

BELOW B-17F, 41-24548, *War Horse,* of the 43rd BG, 403rd BS, crash-landed on the Pacific airfield of Tadji, New Guinea, returning from a supply drop over Hollandia on 4 May 1944. *(USAF)*

Chapter Three

B-17 Combat Mission

Life for Fortress crews during the Second World War was harsh and exhausting. Not only was there the constant threat of fighters and flak, but the cramped, cold, unpressurised interior of the B-17 meant that crewmen needed to wear bulky and uncomfortable flying clothing, and breathe oxygen through a face mask above 10,000ft. For the tail and ball turret gunners, their only link with the rest of the crew was through the intercom system.

OPPOSITE B-17F, DF-Q, during the filming for *Memphis Belle* in 1989. *(Patrick Bunce via Keith Wilson)*

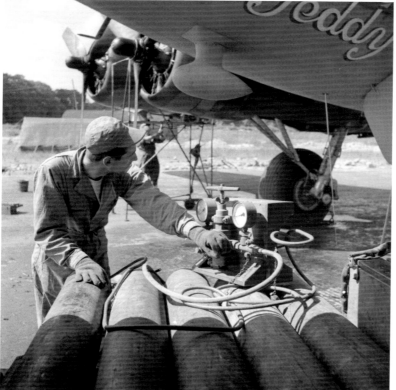

ABOVE A B-17G from the 91st BG in formation high above hostile territory. For many hours the crew had to endure harsh conditions in the unpressurised interior at altitudes of 25,000ft and above. *(ww2 images)*

LEFT Pte A. Steinberg, an ex-waiter from New York, recharges the oxygen tanks of a 92nd BG B-17 at Alconbury for its next mission, June 1943. *(IWM D15110)*

To prepare for an eight to ten-hour mission over Europe, the B-17 crews were woken early, usually before dawn, to attend mission briefings that were followed by an early breakfast. Afterwards they were driven out to their aircraft dispersed around the airfield. The pilots would meet the crew chief responsible for their aircraft and perform a pre-flight inspection with him. Other crew members would attend to their tasks: the gunners would check their weapons and ammunition and the functioning of turrets; the navigator would prepare his equipment and charts, while the bombardier

was responsible for checking that the bombs were correctly loaded and fused, and that the bomb sight was functioning correctly.

Once all inspections and checks were complete, the crew would either man their stations, or in the case of the gunners they would assemble in the radio room in preparation for take-off. Everything from engine start through to taxi, take-off and assembly points was carefully timed. Unlike the British heavy bombers, which normally flew in a loose stream under the cover of darkness, the American heavies grouped together in tight combat boxes and flew as a massed formation. To achieve this, each aircraft was assigned a position within a box, which was itself allocated a position in the formation; each aircraft had then to await its turn to take off at the allotted time. It was common for the maximum weight of the aircraft to be exceeded under combat conditions. Normal maximum was 65,500lb, but loaded with fuel, bombs and ammunition this could be exceeded up to a maximum limit of 72,000lb. Take-off under these conditions was

ABOVE Take-off. A scene from *Memphis Belle* (1989). *(Patrick Bunce via Keith Wilson)*

LEFT Gear up, flaps up. *(Patrick Bunce via Keith Wilson)*

LEFT Climb power set. *(Patrick Bunce via Keith Wilson)*

ABOVE B-17E, 41-9100, *Birmingham Blitzkrieg,* was painted in a highly visible red and white candy-stripe scheme, and flown by the 379th BG as an assembly ship for other BG aircraft to formate on prior to setting off on their mission. *(USAF)*

RIGHT Although the B-17 cockpit is fairly roomy, when the pilots are dressed in their flight clothing movement in and out of the seats is difficult. The pilot sits in the left-hand seat, seen here with his right hand on the throttle levers. *(All photographs of* Mary Alice *by Graeme Douglas)*

FAR RIGHT With his hands on the control yoke, the co-pilot flies the aircraft.

testing, with the aircraft needing the full length of the runway to become airborne. If power was lost on take-off there was a strong chance the aircraft would crash.

Once airborne, each aircraft would climb at a prescribed rate and heading to an assembly point usually located on the East Anglian coast. These points were actually radio beacons and

their positions could be identified even when the ground was obscured by cloud. With so many aircraft in the air at once, assembly could be hazardous when climbing for prolonged periods through cloud. Collisions were not uncommon.

It could take two hours or more orbiting the radio beacon in order to assemble all the

LEFT All four throttle levers can be manipulated with one hand by either the pilot or co-pilot, allowing them to share the workload between flying and controlling engine power settings.

aircraft into their correct positions within the formation and at the briefed combat altitude. At the appointed time, known as 'zero hour' the formation would set course for the Continent. By this time the formation's movements would have been monitored by German fighter controllers.

During the crossing of the North Sea, preparations were made by the crew for combat. Gunners, who up to this time had had little to do, would load ammunition into the breach of their .50-calibre machine guns and test-fire brief bursts. This was an essential check to perform, to ensure that the gun breeches, although heated, were still working at temperatures of −40°C and below. The bombardier would leave his position in the nose of the aircraft and work his way back to the bomb-bay. Breathing oxygen from a small portable bottle, he removed all the arming pins from the bomb fuses. This allowed the fusing mechanism, in the form of a small propeller, to rotate when the bombs dropped, thus arming them as they fell. As the formation approached the enemy coast, crew members would don body armour of special metal-plated vests or 'flak suits'. These consisted of thin steel plates woven into canvas vests and aprons, and although heavy, they had been proved to reduce injuries from shell splinters. Some crew members would also put on metal helmets at

this time, similar to infantry helmets but specially adapted to allow headphones to be worn.

Once the enemy coast was crossed, the crew would be on the alert for hostile aircraft. Everyone was now fully employed. Turret gunners would constantly rotate their turrets to search the sky for enemy fighters. The radio operator had to listen in to coded Morse signals

LEFT A flak vest is worn by the radio operator to protect him from shrapnel.

LEFT The radio operator listens in to the Liaison radio receiver for messages from ground stations.

ABOVE As well as controlling the chin turret the bombardier could operate either of the two cheek guns. These guns were counter-balanced by a cable and spring system.

RIGHT The navigator, who sat at a table at the rear of the nose compartment, could also operate the cheek guns.

RIGHT Early B-17s with their open windows exposed the waist gunners to icy slipstreams. Later series aircraft such as this one had covered and staggered windows which helped, but gunners still needed to wear heavy protective clothing.

from base that might indicate an aborted mission. In the nose, the navigator kept an exact check of the aircraft's position as well as keeping a minute-by-minute report of the flight in the aircraft's log. Even when following other aircraft in formation, it was necessary for him to know his aircraft's exact position. In the event of an emergency, he may be required to plot a course for home. Alongside him the bombardier, until he needed to be ready for the bomb-run, would act as a gunner by operating the chin turret guns.

Radar-guided flak could occur at any time. Seen from the air it appeared as black puffs of smoke and if it was accurate enough, one direct hit in a fuel tank was all that was needed to destroy a Fortress. The crew were helpless against this barrage. Evasion was not possible because the formation had to retain its briefed course and height. Most crew members preferred to have the opportunity, at least, to fight back against enemy fighters.

Encounters with fighters usually occurred when the Allied escorts were reaching the limits of their range. Engagements were normally very brief, the preferred attack method of the Luftwaffe being a downward pass through the bomber formation. Closing speeds were often in excess of 500mph, leaving gunners wielding hand-held weapons only split seconds in which to see, identify, aim and fire as the enemy streaked through. Gunners operating the upper and ball power turrets had a better chance of scoring a hit; their guns were power-driven and they were equipped with computing gun sights, which took into account the forward speed of the aircraft and could calculate the deflection angle required for accurate shooting. The chin turret fitted to the G series aircraft, although not equipped with a computing sight, was an effective deterrent against the head-on attacks preferred by fighters during the early part of the air war over Europe.

When the formation reached the Initial Point (IP) – the position over which the bombers turned to make their run to the target – they were potentially at their most vulnerable. By now the German radar controllers would have figured out exactly where the bomber stream was heading and every anti-aircraft gun in the vicinity would then be brought to bear on the

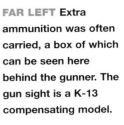

FAR LEFT Extra ammunition was often carried, a box of which can be seen here behind the gunner. The gun sight is a K-13 compensating model.

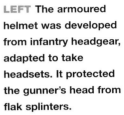

LEFT The armoured helmet was developed from infantry headgear, adapted to take headsets. It protected the gunner's head from flak splinters.

FAR LEFT To fit inside the ball turret the gunner had to be a small man. Despite the turret's vulnerable position he was well protected by armour plate. There was no room for a parachute inside, so if the aircraft spun out of control his chances of escape were slim. *(ww2 images)*

LEFT When manning his turret behind the pilots' seats, the engineer and the turret structure effectively block access through to the bomb-bay.

FAR LEFT The tail gunner knelt, squeezed between two ammunition boxes. Apart from the intercom he was isolated from the rest of the crew.

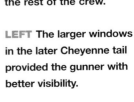

LEFT The larger windows in the later Cheyenne tail provided the gunner with better visibility.

attackers. At this time enemy fighters would disperse, not wishing to become victims of their own flak barrage. It became standard practice in the latter stages of the war that only the lead aircraft of each squadron carried a bomb sight; the following aircraft dropped their bombs on the leader's signal. If the target was obscured by cloud, the role of lead aircraft fell to the so-called 'Mickey Ship' that carried H2X radar, which allowed the formation to bomb through overcast. On the run-in from the IP to the target, pilots still needed to fly a tight formation despite often coming under the most intense flak. It was important that the aircraft maintained a close grouping in order that accurate bombing results could be obtained. Crews often felt at their most vulnerable at this stage of the mission. At the required point, the bombardier opened the bomb doors and then at the exact moment that the lead aircraft dropped its load, he would manually release his bombs. Upon release, the aircraft would gain some altitude and an increase in airspeed. Bomb doors would be closed and the formation would turn on to a rally point before setting course for the journey home.

The return flight would follow a different route to the outbound leg in order to evade the defences already alerted. Routes were planned to avoid the worst of known flak concentrations. Again, vigilance was required from all crew members as they left the target area; they knew the attacking fighters would soon return. The Luftwaffe would pick out stragglers, aircraft that had sustained battle damage or were suffering mechanical difficulties and subsequently unable to keep up with the formation.

Damaged aircraft would sometimes try to head for a neutral country such as Switzerland or Sweden if it appeared that their chances of making it back to England were slim. Others, while attempting to make their home base would end up either diverting to the first airfield they encountered on crossing the English coast, or in the worst situations, ditching in the Channel. Even this was thought to be preferable to a forced landing in enemy territory, with the prospect of the crew spending the remainder of the war in a POW camp.

In a mission that could last up to nine or ten hours, creature comforts on board the Fortress were few. A basic chemical toilet was provided, along with a number of 'relief tubes'. The difficulties encountered in using these facilities in freezing temperatures can probably be imagined. It was said that if you needed to use the relief tube, you should make sure you were the first to do so, for after the first use it always became blocked! Food rations and Thermos flasks of coffee were carried, but eating and drinking was best done below 10,000ft when oxygen masks were not required.

Once the bombers crossed the enemy coast on their return, the formation would gradually loose height and the crew could start to relax a little. Enemy fighters were unlikely to attack at this point, but the gunners still needed to remain vigilant. While crossing the Channel, the formation usually descended below 10,000ft and the crew were then relieved of the need to breathe oxygen through a face mask which, after a long mission, was a great relief. As the English coast was crossed, gunners would remove the ammunition from their guns and leave their turrets and make their way to the radio room in preparation for landing. This was considered to be the safest part of the aircraft to be in the event of a crash-landing; those crew members without a seat would brace themselves against the bulkheads.

On reaching their home base, the Fortresses would peel out of formation in preparation for landing. Priority was given to any aircraft with battle damage or with wounded men on board. Pilots needed to maintain their concentration at this time. With so many aircraft orbiting the airfield, tiredness could lead to misjudgements and collisions from which there was almost certainly no chance of survival.

Once safely down on the ground the crew could look forward to a meal, but not before a debriefing with the intelligence officers and the filing of combat reports. As for the ground crew, their work would begin anew with a thorough inspection of the bomber for any damage and defects. If there was any battle damage it would often require the ground crew to work through the night so that the aircraft was serviceable for a repeat mission the next day.

B-17 pilot

Darrell Blizzard was a B-17 pilot in the 535th Bomb Squadron, 381st Bomb Group, based at Ridgewell in Essex. He recalls joining his group:

'My crew was one of the last to join the 381st on Easter Sunday 1945; consequently we got only five combat missions before VE-Day.'

Despite the air war in Europe coming to a close, with the defeat of Germany only a month away, there was no shortage of dangerous incidents, sometimes unrelated to enemy action. Darrell remembers:

'When we first arrived at the base we were assigned to fly an older B-17, which had been shot up pretty badly. We took it on a check ride after some extensive repairs had been made. No sooner had we broken ground on take-off than the cockpit filled completely with gas [fuel] fumes to the point that we had to open both windows to prevent asphyxiation! I immediately made a 360° turn and set that thing down so smoothly that the crew didn't know we were on the ground! I was not about to cause any movement that might produce a spark! We did not find out until later that no one else would fly this aircraft for the same reason that we would never fly it again.'

The hazardous nature of combat operations is well illustrated by a mission that Darrell Blizzard describes which, through a series of incidents, almost led to tragedy at the very end of a long mission:

'Another occurrence happened on our third mission when the groups were making extensive use of chaff – the aluminium strips that were dispersed to try to mess up the accuracy of the flak gunners on the ground. We were flying in a lower element of the group and as the chaff drifted through our element, a piece of it swept into our open bomb-bay and got entangled in the worm gears that controlled the bomb doors. Consequently, we were unable to close the doors electrically after the bombs were dropped, and had to send the engineer back with an oxygen bottle to crank them up manually. This took some 15 minutes to accomplish and all the excess drag caused us to lag far behind the rest of the group, even using full power.

'We finally caught up to the rest of the group just as we got to the Channel, but by that time [this was a ten-hour flight] we were just about out of gas. I was flying co-pilot on that mission and was expecting the pilot to issue a Mayday to the base and fly straight in, since we had already lost one engine. He kept flying formation even though we were the second element to enter the traffic pattern for landing. Finally we got into the pattern and on the downwind leg we lost the second engine, but he continued as though nothing had happened … while I'm squirming in my seat! On the base leg, just before making our turn on to the approach, we lost the third engine, but fortunately had enough momentum to make the landing, pull off to the edge of the runway and sit there! I never flew with him again! But it did prove that the B-17 would just about fly on fumes!'

BELOW The Blizzard crew, pictured outside their crew hut at Ridgewell. Darrell Blizzard is second from the left in the back row. *(Darrell Blizzard)*

Chapter Four

Anatomy of the B-17

───●────────────────────

When the last Fortress was completed more than sixty years ago, who then would have thought examples would still be flying today? The average life expectancy of a B-17 in wartime was measured in weeks. The tens of thousands of hours spent on design and construction could vanish in a split second with the explosion of a flak shell or a burst of cannon fire. Yet, the big Boeing was designed with longevity in mind.

OPPOSITE A brand new B-17G in natural metal finish, illustrating the olive drab anti-dazzle paint applied to the nacelles and forward of the cockpit windscreen. The walkways and fuel filler caps are visible on the wings. *(IWM NYP14211)*

B-17G Flying Fortress

(Mike Badrocke)

1 Right elevator rib structure
2 Tailplane leading edge de-icing boot
3 Two-spar tailplane torsion box structure
4 Leading edge ribs
5 Fin and tailplane mounting bulkheads
6 Elevator hinge control, cable actuated
7 Rudder hinge control, cable actuated
8 Ammunition boxes, 565 rounds each
9 Elevator trim tab
10 Tail gunner's armoured protection panel
11 Tail gun turret, 2 x M-2 Browning 0.5in machine guns (mg)
12 Gun sight
13 Rudder tab
14 Fabric-covered rudder rib structure
15 Fin rib structure
16 HF antenna cables
17 Fin leading edge de-icing boot
18 Left fabric-covered elevator
19 Left tailplane
20 Fin root
21 Tailwheel retraction screw jack
22 Tailwheel housing
23 Shock absorber strut
24 Retractable castoring tailwheel
25 Entry door
26 Crew toilet
27 Rear fuselage frame and stringer structure
28 Auxiliary DC generator
29 Hand fire extinguisher
30 Marker beacon antenna cable
31 Fuselage walkway
32 Gunner's footstep, left and right
33 Ammunition boxes, 600 rounds each
34 Demand oxygen regulators
35 Antenna mast
36 Left waist gunner's station, 0.5in mg
37 Right waist gunner's station, 0.5in mg
38 Portable oxygen bottles
39 Trailing edge antenna winch
40 Lower turret support structure
41 Rotating Sperry ball turret, 2 x 0.5in mg with 500rpg
42 Radio equipment racks
43 Ammunition box, with 300 rounds
44 Radio operator's flexibly mounted 0.5in mg
45 Radio compartment's glazed jettisonable roof hatch.
46 Radio operator's station
47 Ventral camera compartment
48 Auxiliary crew member's seat
49 Under-floor propeller de-icing fluid tank
50 Wing fixed trailing edge rib structure
51 Cooling air spill ducts
52 Flap shroud ribs
53 Split trailing edge flap
54 Right aileron rib structure
55 Aileron mass balance weight
56 Aileron shroud ribs
57 Wing tip section joint rib
58 Wing tip rib structure

59 Right navigation light
60 Leading edge pneumatic de-icing boot
61 Outer wing panel built-up rib structure
62 Outboard Tokyo tanks Nos 1–5, total capacity 270 US gal
63 Outer wing panel spar joint
64 Outboard Tokyo tanks Nos 6–9, total capacity 270 US gal
65 Right landing light
66 Engine supercharger ram air intake
67 Intercooler intake
68 Exhaust driven supercharger
69 Supercharger cooling air intake
70 Engine oil tank

71 Fireproof engine mounting bulkhead
72 Engine bearer struts
73 Engine mounting ring frame
74 Nacelle cowl flaps
75 No. 4 engine nacelle
76 Detachable engine cowling panels
77 Supercharger pressure duct to carburettor
78 Intercooler
79 Air filter
80 Nacelle access panel
81 Leading edge rib structure
82 No. 4 engine fuel tank, capacity 425 US gal
83 No. 3 engine feeder tank, capacity 212 US gal
84 Inboard engine intercooler
85 Main wheel leg pivot mounting
86 Oil cooler intake
87 Exhaust duct to inboard engine supercharger
88 Main wheel leg strut
89 Main wheel with 56in tyre
90 Drag strut
91 No. 3 engine nacelle
92 Wright R-1820-97 Cyclone 9-cylinder radial engine
93 Exhaust system
94 Engine accessory equipment
95 Main wheel retraction screw jack
96 Main wheel bay
97 Inboard engine oil tank
98 No. 3 engine fuel tank, capacity 213 US gal
99 Bomb bay, max capacity 10,000lb
100 Vertical bomb racks, left and right
101 Bomb bay central catwalk
102 Rope handrail
103 Dinghy stowage
104 Left flap shroud ribs
105 Trailing edge skin panelling
106 Left split trailing edge flap
107 Cooling air spill ducts
108 Wing flap actuating strut
109 Aileron tab, left only

110 Left fabric-covered aileron
111 Aileron mass balance weight and cable-operated hinge link
112 Wing panel corrugated inner skin panel
113 Outer skin panelling
114 Left navigation light
115 Leading edge de-icing boot
116 Left outboard Tokyo tanks
117 Landing light
118 No. 1 engine nacelle
119 Oil cooler ram air intake
120 No. 1 engine fuel tank
121 No. 2 engine feeder tank
122 Inboard wing panel skin panelling
123 Sperry upper turret, 2 x 0.5in mgs, with 400rpg
124 Cockpit bulkhead and doorway
125 Hydraulic reservoir and accumulator
126 Thermos bottles
127 Emergency hydraulic accumulator
128 Turret foot platform
129 Oxygen bottles
130 Access hatch to nose compartment
131 Control cable runs
132 D/F loop antenna
133 Under-floor equipment bay
134 Forward entrance hatch
135 Rudder pedals
136 Control column and hand wheel

137 Co-pilot's seat
138 Pilot's seat
139 Cockpit roof glazing
140 Pilot's radio controllers
141 Instrument panel
142 Windscreen panels
143 No. 2 engine nacelle
144 Hamilton Standard 3-blade hydromatic constant speed full feathering propellers
145 Propeller hub pitch change mechanism
146 Localiser antenna
147 Astrodome
148 Navigator's table
149 Right cheek gun
150 Portable oxygen bottles
151 Pitot head
152 Ammunition boxes, left and right cheek guns
153 Bombardier's station
154 Nose compartment left 0.5in gun position
155 Chin turret reflector sight
156 Norden bomb sight
157 Frameless Plexiglass nose fairing

158 Optically flat bomb sight panel
159 Bendix chin turret, 2 x 0.5in mgs, with 365rpg
160 300lb HE bomb
161 500lb HE bomb
162 1,000lb HE bomb
163 2,000lb HE bomb

68

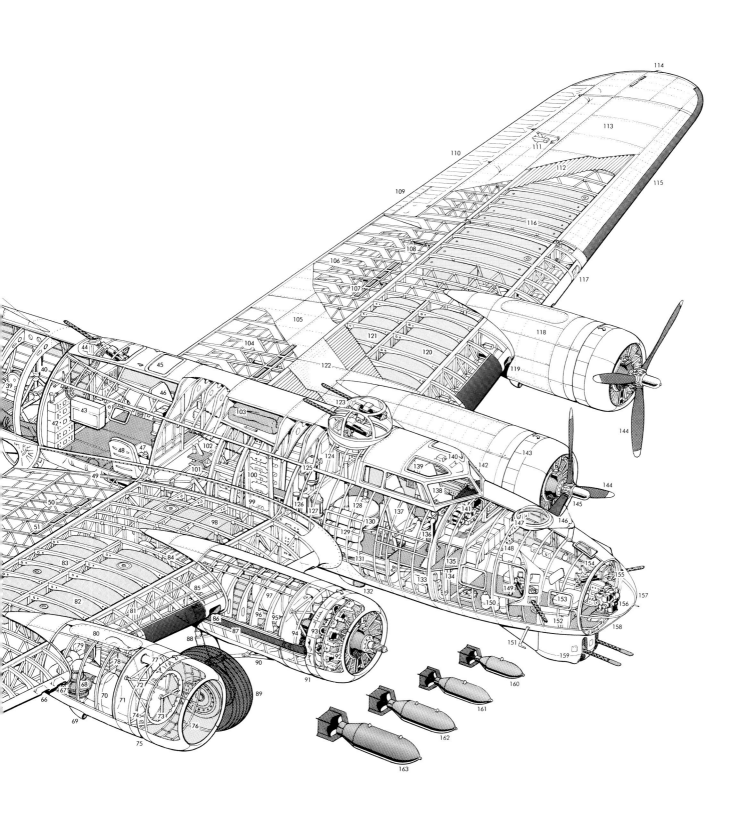

**General arrangement
drawing of the B-17G.**
(USAF)

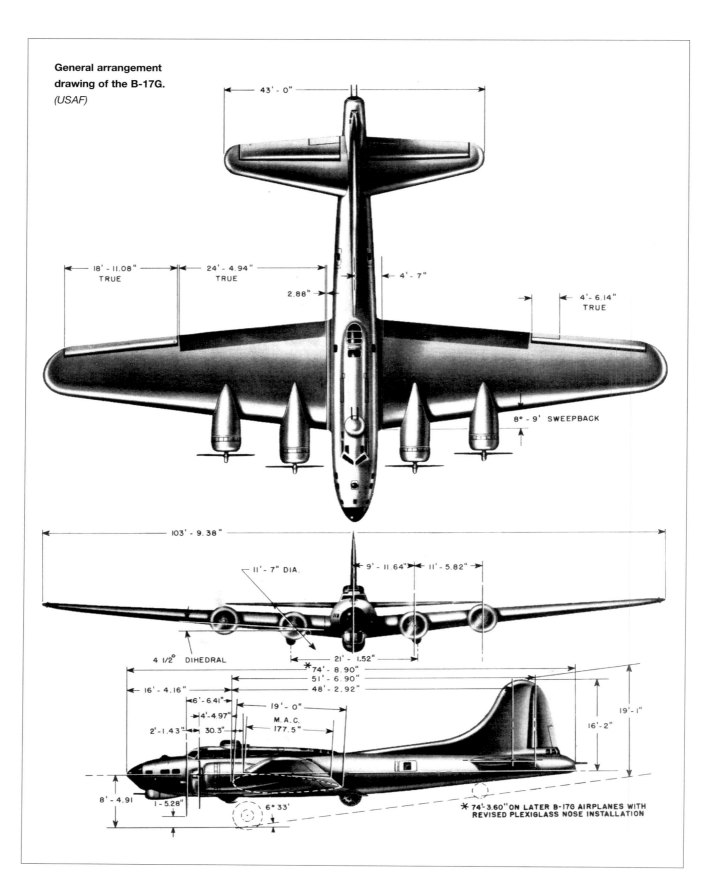

43' - 0"

18' - 11.08" TRUE

24' - 4.94" TRUE

2.88"

4' - 7"

4' - 6.14" TRUE

8° - 9' SWEEPBACK

103' - 9.38"

11' - 7" DIA.

9' - 11.64"

11' - 5.82"

4 1/2° DIHEDRAL

21' - 1.52"

* 74' - 8.90"

51' - 6.90"

48' - 2.92"

16' - 4.16"

6' - 6.41"

19' - 0"

19' - 1"

4' - 4.97"

M. A. C. 177.5"

16' - 2"

2' - 1.43"

30.3"

8' - 4.91"

1 - 5.28"

6° 33'

* 74' - 3.60" ON LATER B-17G AIRPLANES WITH REVISED PLEXIGLASS NOSE INSTALLATION

Fuselage

The following description is typical of a late production B-17G. Where there are significant production variations compared with earlier versions these are highlighted in the text.

The all-metal fuselage is a semi-monocoque design built of circumferential stiffeners or stations, and extruded longitudinal stringers and longerons to which aluminium alloy stressed skin is attached by rivets. Stations are numbered from 1 in the nose to 11 at the tail with sub-stations allocated letters, eg, 6-D. At stations 3, 4, 5, 6 and 9 structural bulkheads are fitted to act as partitions and add strength to the structure. Carrying the loads between the wings are steel chords, and truss-type compression struts on either side of the bomb-bay take the load between the spars. Running longitudinally through the bomb-bay is another truss structure, which takes the load of the inboard bomb racks; it also serves as a walkway through this part of the aircraft.

During manufacture the fuselage is constructed in a number of sub-assemblies consisting of the nose section, cockpit, bomb-bay, radio room, waist and tail gunner's compartment. For field repairs the only recommended fuselage split is at station 6 – the joint between the radio room and waist; and also at station 11 – the tail gunner's compartment. Three large section longerons run from stations 1 to 4 and carry the major

loads; an additional four similar longerons run between stations 4 to 6-D. Numerous smaller section stringers form the structure from nose to tail and are attached to the stations at regular intervals by brackets secured by rivets.

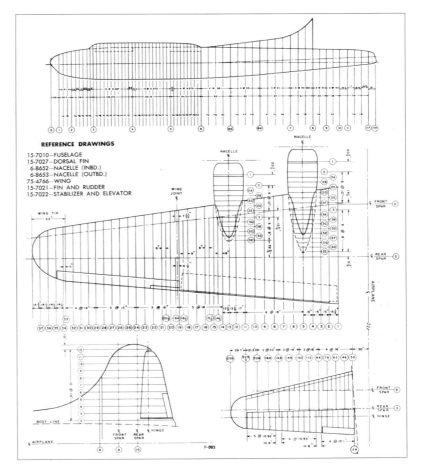

ABOVE Fuselage stations and rib numbering are illustrated in this drawing. *(USAF)*

FAR LEFT Assembly of the first formers and stringers on the front left section of the fuselage. *(Lockheed)*

LEFT The skin is riveted to the stringers using a pneumatic riveter. *(Lockheed)*

Nose section

This section provides crew accommodation for the navigator and bombardier. The navigator sits at a work table to the left side; the bombardier sits on a small seat at the front of the nose with the bomb control and release panel to his left. To his right, when installed, are the ammunition boxes for the cheek guns. Bolted to station 1 is the clear Plexiglass

nosecone which contains an optically flat panel for sighting the Norden bomb sight. Armament carried in the nose depends on the variant: E and early F series Fortresses carried up to six ball-and-socket mounts in the nosecone and side windows for .30-calibre machine guns. Later F series aircraft had the enlarged staggered windows on either side of the nose carrying .50-calibre guns. The final G series had staggered cheek windows, but with the stagger reversed and with a remotely controlled 'chin' turret housing two more .50-calibre guns installed below the bombardier's seat. This unit is operated by the bombardier using a swing-out controller arm fixed to the floor of the nose. Entry to the nose compartment is via the crawl way below the pilot's compartment, which is accessed either from the cockpit, or from outside using the nose entry hatch.

Pilots' compartment

Situated behind the nose, between stations 3 and 4, this area contains the cockpit in which the pilot sits in the left-hand seat and the co-pilot to the right. Behind, the flight engineer/top gunner occupies the upper gun turret.

The two pilots each operate a control column consisting of a three-spoke semi-circular yoke mounted on the vertical column moving fore and aft. The main flight instruments

can be viewed by both pilots as they are in the middle of the panel; directly in front of the pilot are a variety of instruments such as the radio compass, voltmeter, suction gauge and oxygen pressure and flow gauges. To his left-hand side he has the main electrical services panel containing switches and ammeters for the engine-driven generators, switches for the three aircraft batteries, for the hydraulic pump and external lights. Further back on the left of his seat, the aileron trim can be adjusted by a small hand wheel. In front of the co-pilot are grouped the main engine instruments – manifold pressure gauges, tachometers, fuel pressure gauges, oil pressure and oil temperature gauges, along with the cylinder head temperature and carburettor air temperature indicators. Each of these gauges has a dual needle indicator for two engines marked either as '1' and '2' on one gauge, and '3' and '4' on the other or sometimes, 'L' and 'R'. A sub-panel to the co-pilot's right contains indicator lights for the carburettor air filters, engine starter switches and the fire extinguisher control. Lower down on the right-hand sidewall an emergency hand-pump for the hydraulic system is installed, behind the engine fuel primer hand-pump in the floor. Just forward of this, a large floor-mounted box houses the four levers for controlling the intercooler air.

Between the two pilots the central pedestal

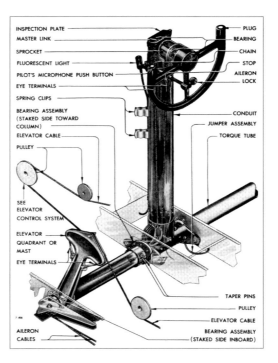

assembly contains controls that can be operated by both pilots – the four throttle levers, propeller pitch levers, the mixture levers and the turbo boost controls. At the front of the unit is the autopilot (AFCE) control box, adjacent to which is a large elevator trim wheel. On the floor just in front of the pedestal, another large wheel controls the rudder trim; just behind this are two locking levers, one for the flight controls and the other for the tail wheel. The flight controls are

LEFT The pilot's control column is linked to that of the co-pilot via the torque tube. The two spring-clips hold the aileron lock in flight. *(USAF)*

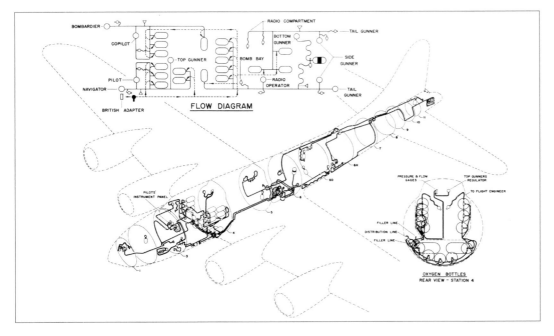

LEFT Oxygen system layout showing 18 bottles with 4 distribution circuits. *(USAF)*

unlocked when the lever is down and the tail wheel is locked when down. At the top of the pedestal another electrical switch box contains controls for the magnetos, flaps, landing gear, identification lights and landing lamps. Four hydraulic control valves regulate the opening of the engine cowl flaps.

The Sperry upper turret is located directly behind the pilots' seats. The gunner operates his turret by adjusting two footrests mounted on the support posts, stepping into the structure so that his head is level with the gunsight inside the turret dome. Early turrets carried their own separate oxygen supply in a tank strapped to the turret framework; later models are plumbed in to the aircraft's oxygen system. Electrical power is supplied through the base via slip rings and brushes and the gunner has an electrical switch box to control items such as gun heaters, ammunition boosters, lead lamp and the gun sight. On each side of the turret,

The pilots' instrument panel of the Imperial War Museum's *Mary Alice*.

1 Localiser indicator	7 Rate of climb indicator	11 Oxygen flow and pressure gauges	16 Oil temperature gauges	20 Oil dilution switches
2 Directional gyro	8 Flux gate compass indicator	12 Manifold pressure gauges	17 Cylinder head temperature gauges	21 Parking brake
3 Flight indicator	9 Pilot's directional indicator	13 Tachometers	18 Outside air temperature gauge	22 Propeller pitch levers
4 Altimeter	10 Suction gauge	14 Fuel pressure gauges	19 Carburettor air filter indicators	23 Lock for pitch levers
5 Air Speed Indicator		15 Oil pressure gauges		24 Pilot's control column
6 Turn and bank indicator				25 Co-pilot's control column

attached to the sidewalls are mounted two groups of oxygen cylinders, five on the left and three on the right. Each pressurised cylinder contains enough oxygen for 4½ man hours' duration at 30,000ft. On the right-hand sidewall under the oxygen cylinders, the hydraulic service panel is installed, comprising the electric motor, pump, filter and associated tubing. The pump supplies pressure to the hydraulic accumulator on the forward face of station 4. Next to the accumulator, a hydraulic reservoir stores sufficient fluid for the system, which supplies the wheel brakes and the engine cowl flaps. On the opposite side of the doorway, on the left-hand side of station 4, is the main electrical fuse panel through which power is distributed to most of the aircraft's electrical services.

The area below the pilots' compartment floor is accessed via the hatchway between the two pilots' seats. It contains engine fire extinguisher bottles, an autopilot gyro unit and

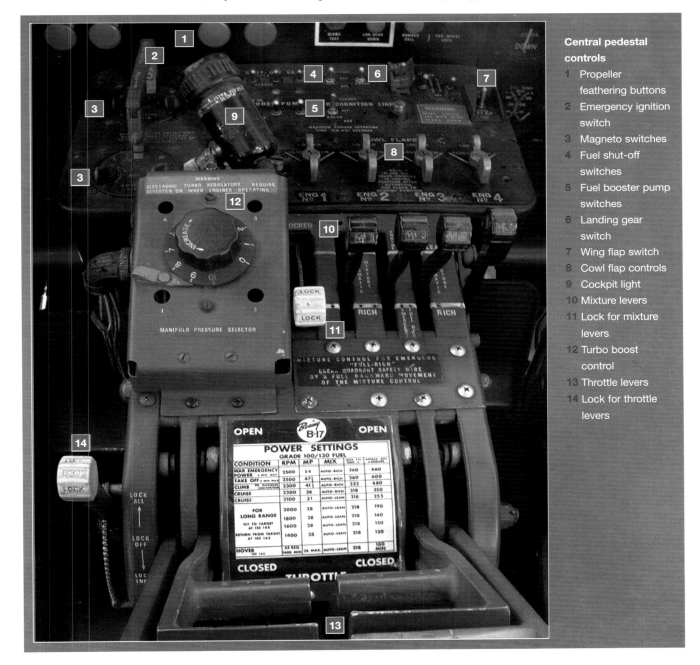

Central pedestal controls

1 Propeller feathering buttons
2 Emergency ignition switch
3 Magneto switches
4 Fuel shut-off switches
5 Fuel booster pump switches
6 Landing gear switch
7 Wing flap switch
8 Cowl flap controls
9 Cockpit light
10 Mixture levers
11 Lock for mixture levers
12 Turbo boost control
13 Throttle levers
14 Lock for throttle levers

ABOVE **The right-hand panel contains the engine instruments.**

BELOW **Left-hand instrument panel.**

seven more oxygen bottles. The nose hatch is on the left-hand side and can be released in an emergency by a quick-release handle that jettisons the hatch.

Bomb-bay

Immediately aft of the pilots' compartment, occupying the space between stations 4 and 5, which form the attachment points for the two wing spars, is the bomb-bay. A catwalk provides access through the centre, between the two inboard bomb racks. Outboard racks are fitted to the fuselage sidewalls. Bombs are attached to shackles clipped to the racks and suspended horizontally. The maximum size of bomb that can be accommodated is 2,000lb on each side, due to the length restriction of the bomb-bay. There is provision to install extra fuel tanks in place of bombs to increase the range of the aircraft. The two bomb doors are hinged from the fuselage sides and are electrically

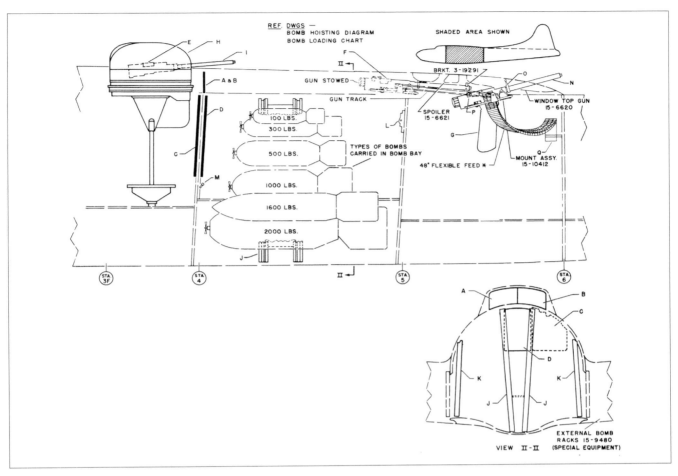

REF. DWGS —
BOMB HOISTING DIAGRAM
BOMB LOADING CHART

SHADED AREA SHOWN

GUN STOWED

GUN TRACK

BRKT. 3-19291

100 LBS.
300 LBS.
500 LBS.
1000 LBS.
1600 LBS.
2000 LBS.

TYPES OF BOMBS
CARRIED IN BOMB BAY

SPOILER
15-6621

48° FLEXIBLE FEED *

WINDOW TOP GUN
15-6620

MOUNT ASSY.
15-10412

STA
3F
STA
4
STA
5
STA
6

VIEW II-II

EXTERNAL BOMB
RACKS 15-9480
(SPECIAL EQUIPMENT)

ABOVE The fuselage centre section offensive and defensive armaments are shown in this illustration. All bombs were carried horizontally, the largest size being the 2,000lb. *(USAF)*

LEFT 500lb bombs are suspended from the racks. The shackles at the bottom of the racks are for the larger 2,000lb bombs.

FAR LEFT Two bomb winches and handles and a winching cradle are stowed on the rear bomb-bay bulkhead.

Radio equipment

The original radio equipment fit varied in detail. This list is typical for a late production B-17G.

SCR 274-N Command radio system – mounted in two racks in the radio room and comprising two transmitters and three receivers. It was used for short- to medium-range air-to-air and air-to-ground communications.

SCR-287-A Liaison radio system – mounted on shelves and brackets in the radio room. The system consisted of a single transmitter with up to six plug-in tuning units and separate table-mounted receiver. The set was used for sending long-range Morse code communications to ground stations.

SCR-269-G Radio compass – receiver mounted in the crawl way below the pilots' seats, with an antenna mounted in a streamlined housing on the lower fuselage forward of the bomb-bay doors, and a whip antenna just behind the chin turret. The system could be tuned in to radio beacons and was used for direction finding.

RC-43 Marker beacon radio – the receiver mounted in the waist compartment was close to the rear entry door. The antenna was fitted below the fuselage. It provided a visual cue when detecting navigational beacons.

SCR-695 IFF receiver/transmitter – mounted on a rack in the waist compartment on the forward bulkhead. It was used to identify the aircraft as friendly to Allied radar. The unit was equipped with a detonator circuit to destroy it if the aircraft forced-landed in enemy territory.

RC-36 Interphone system – had a dynamotor and amplifier mounted in the radio room and jack boxes at every crew station, allowing crew communication using throat microphones and headsets.

SCR-522-A – this VHF set had the transmitter and receiver mounted under the floor of the radio room, a switch box in the cockpit and an antenna mounted at the forward end of the dorsal fin; it was used for voice communications between bombers and fighters.

SCR-570 Landing approach radio – comprised of two receivers and dynamotors mounted on the floor behind the co-pilot's seat, a cockpit-mounted control box and indicator and a distinctive antennae assembly mounted just ahead of the astrodome. The system provided the pilot with both vertical and lateral indication of his approach to land in poor visibility.

SCR-578 Emergency transmitter – for use in the event of a ditching, was mounted in a yellow flotation bag and included either a kite or hydrogen-inflated balloon to hoist the antenna aloft at sea. The unit was stowed close to the ball turret or in the radio room.

BC-221 Portable frequency meter – mounted in a box in the radio room.

driven by a motor under the forward end of the catwalk. Just above the catwalk the fuel transfer pump is situated on the aft side of station 4, which transfers fuel between the wing tanks.

Radio compartment

Situated between stations 5 and 6, this compartment has a seat on the left-hand side for the radio operator, who sits at a small table in front of the Liaison receiver. Additional seating is sometimes supplied for other crew members. There is a hatch in the floor giving access to a camera well containing the strike camera and viewfinder. Also stored under the floor are three more oxygen cylinders. Most B-17s have a .50-calibre gun firing aft through the roof hatch. In early versions, the Plexiglass panel is removed and the gun, mounted on a sliding yoke, then swung out of the opening. Later aircraft have the gun mounted in a swivelling mount through a cut-out in the panel, allowing the panel to stay in place at all times. Radio equipment is mounted on the bulkheads in racks, or is floor-mounted. Emergency hand cranks and extension shafts for lowering the landing gear, flaps or bomb doors, are clipped to forward side of station 6.

Rear fuselage (waist) compartment

This is the area between stations 6 and 11. The Sperry lower (or ball) turret is hung from a reinforced box structure between frames 6A and 6B. The turret is suspended from a tube and frame assembly, allowing 360° of rotation horizontally inside a circular ring gear mounted in the floor of the aircraft; it can also rotate 90° vertically. Early turrets carried their own oxygen supply on the frame assembly, although in later models the turret is plumbed into the aircraft's own supply. Electrical power is supplied via slip rings and brushes. Entry to the turret can be achieved only when the aircraft is in flight. The gunner hand cranks it into position and then unlatches the door before stepping down into the turret and closing the door behind him, which forms a back rest. He is seated on a padded steel plate with his legs bent and raised

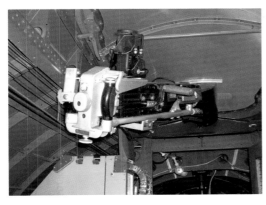

FAR LEFT The radio operator's gun fired to the rear. Ammunition is held in the wooden box below the gun and fed to the gun using the flexible chute, which is seen emerging from the top of the box.

LEFT The radio operator's chair and table are positioned on the left-hand side of the radio compartment.

above him so that he is curled up in an embryo-like position. On each side of him are the two .50-calibre guns. Movement is achieved by actuating two handgrips in front of him. His face is level with the eyepiece of the computing K-4 gun sight and his left foot actuates the range control input for the sight. Power is provided by the same electro-hydraulic unit as used in the Sperry upper turret.

A wooden walkway allows crew access

around the turret to the rear of the aircraft. Further down the waist compartment the two side gunners each man a .50-calibre weapon. The gunners stand in front of large windows giving them good visibility to the sides of the aircraft. Wooden ammunition boxes are mounted on the floor or against the fuselage sides. The design of the windows and gun mountings varies considerably – an early design had lift-out panels in the side windows,

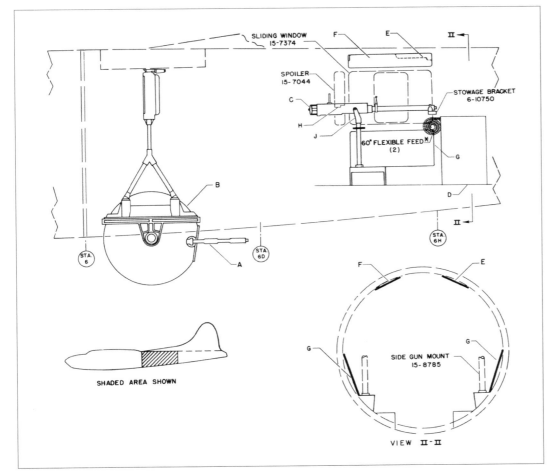

LEFT The armament diagram for the waist shows the position of the ball turret and the early style of post-mounted machine guns with sliding windows. *(USAF)*

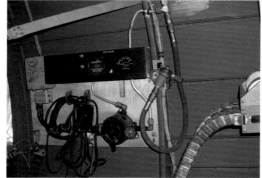

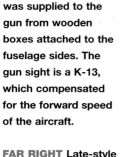

which had to be removed before the post-mounted guns were swung out into the firing position. Later designs had the guns installed on the window sill with the window glazed with

Plexiglass. Various designs of gun sights were used, from early ring and bead types through reflector sights to fully compensating sights on the final versions.

At the rear of the compartment, close to the tail wheel, the crew entry door opens outwards. Like all access hatches, this has quick-release pins to jettison the door in an emergency. To the rear of the door behind station 7 is the retractable tail wheel. The oleo strut retracts upwards and forwards towards the fuselage roof when the undercarriage is raised. Behind the oleo is the electric retraction motor and drive screw. There is just room for the rear gunner encumbered in heavy flight clothing to crawl past the tail wheel to access his compartment in flight.

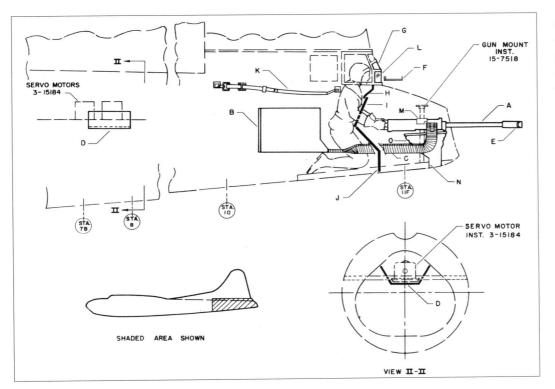

SERVO MOTORS
3- 15184

SHADED AREA SHOWN

SERVO MOTOR
INST. 3-15184

VIEW II-II

Tail gunner's compartment

Aft of station 11 is positioned the tail gunner's compartment, housing two .50-calibre guns, ammunition boxes and the gunner's seat. The gunner sits on the padded bicycle-style seat and rests his lower legs on two padded knee supports. This squatting position places the handles of the machine gun mounts level with the gunner's chest. Early B-17s had the original 'stinger' tail compartment with a limited range of movement for the guns and a ring and bead sighting system. The improved 'Cheyenne' turret provided the gunner with more room and a greater range

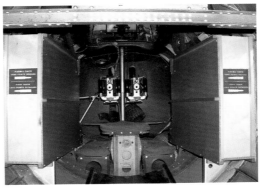

of movement for his guns, as well as the installation of an N-7 reflector sight. Behind him is an emergency escape hatch positioned below the right elevator.

LEFT The tail gunner sat on the small saddle with his knees under him in a semi-kneeling position. Ammunition feed to the guns was through flexible chutes. This is the 'Cheyenne' tail.

FAR LEFT A Vega factory worker prepares the tail gunner's access hatch for fitting to a B-17F. *(Lockheed)*

LEFT The early stinger-style tail shown here on *The Pink Lady* had a simple ring and bead sight.

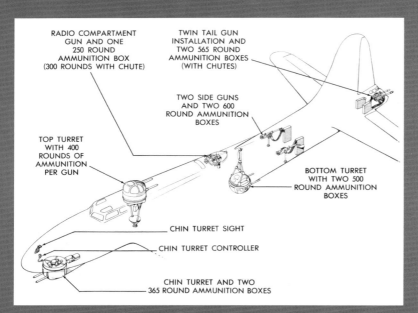

RADIO COMPARTMENT
GUN AND ONE
250 ROUND
AMMUNITION BOX
(300 ROUNDS WITH CHUTE)

TWIN TAIL GUN
INSTALLATION AND
TWO 565 ROUND
AMMUNITION BOXES
(WITH CHUTES)

TWO SIDE GUNS
AND TWO 600
ROUND AMMUNITION
BOXES

TOP TURRET
WITH 400
ROUNDS OF
AMMUNITION
PER GUN

BOTTOM TURRET
WITH TWO 500
ROUND AMMUNITION
BOXES

CHIN TURRET SIGHT

CHIN TURRET CONTROLLER

CHIN TURRET AND TWO
365 ROUND AMMUNITION BOXES

ABOVE The B-17G is defended by three power-operated turrets, a tail gun position, and up to five flexible machine guns. The two cheek guns are not shown in this illustration. The quantities of ammunition shown are those recommended; gunners would often carry extra supplies. *(USAF)*

Improving the breed – defensive armament

The Model 299, despite being dubbed a Flying Fortress in 1935, was a fortress in name only and was far from being able to defend itself with a complement of five .30 or .50-calibre guns. None of these weapons were mounted in power turrets, but instead were housed in glazed gun blisters with heavy framing and an extremely limited range of movement, or in a manually operated nose

cupola. Small improvements followed with the C and D series: the belly blister was replaced by a larger 'bathtub' giving the gunner more space and a little more range of movement. Even so, these aircraft were still woefully armed when compared to the new breed of fighters equipped with machine guns and 20mm cannon, as the RAF discovered when they took the B-17C into combat in 1941 as the Fortress I.

To their credit, Boeing recognised the limitations of the original design of what was essentially a maritime patrol aircraft and which was now expected to fulfil the role of a strategic bomber. They responded rapidly with a redesign of the airframe to include the much needed power operated turrets. In the early war years the British had developed their own turret designs – the Fraser Nash and Boulton Paul designs – but because of America's neutrality at the time they refused to share the technology, fearful that it might find its way into the hands of their enemies. So with a clean sheet of paper, Sperry began design work on the turrets that would defend the Fortress from above and below.

Sperry A-13 and A-13A ball turrets

With the E series, the B-17 now had two power-operated turrets and a manned twin-gun installation at the rear. The original belly turret on the first 112 B-17Es was a remotely sighted Sperry unit, installed as a stopgap until the manned Sperry ball turret was available. The new ball turret was a big leap forward in the defence of the bomber from attacks from the underside. The gunner with his communication, heating and oxygen connections, machine guns and gun sight, and the turret-operating controls and electro-hydraulic power unit, were all contained in a sphere of only 44in diameter. The turret could rotate through 360° in azimuth, and between 22° and 90° in depression. There

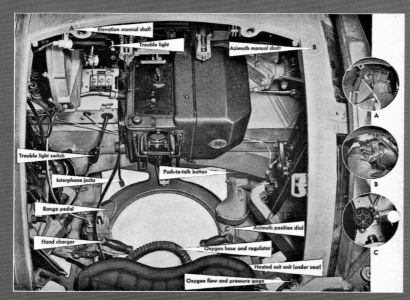

Elevation manual shaft

Trouble light

Azimuth manual shaft

Trouble light switch

Interphone jacks

Push-to-talk button

Range pedal

Azimuth position dial

Hand charger

Oxygen hose and regulator

Heated suit unit (under seat)

Oxygen flow and pressure gage

LEFT Looking into the Sperry ball turret, this is the view that the gunner had as he stepped in. The seat is at the bottom of the picture, and a foot rest on either side of the circular window. *(USAF)*

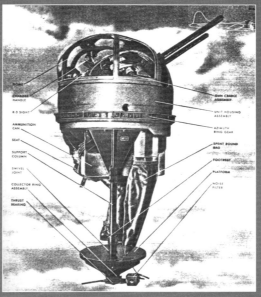

was a cut-off cam incorporated in the gun-firing mechanism to prevent the machine guns from firing through the propeller arcs. At the heart of the turret was the K-4 computing gun sight. It took movement information from the drive mechanism of the turret and computed the point of aim based upon the direction and rate of rotation and angle of depression, the size of the attacking fighter and its range. The gunner was able to frame the attacker in the reticle of his sight by applying pressure to a switch under his left foot. The Sperry ball was the only really effective manned under-turret in use during the Second World War.

Sperry A-1, A-1A and A-1B upper turrets

To provide defensive protection to the aircraft from attacks from above, the upper turret was mounted immediately behind the pilots' seats; it could rotate a full 360° in azimuth and from 0° to 85° in elevation. A fire cut-off cam prevented the guns from firing through the propeller arcs or the tail. The turret was manned by the flight engineer who stood on adjustable foot rests and used a sling-type seat that he could adjust to bring his eyes level with optical head of the K-3 gun sight. The turret housing contained the electro-hydraulic power units, drive transmissions and fire cut-off cam, along with the two .50-calibre machine guns. The gunner operated the turret using a twin-handled controller. By moving

the controller to the left or right, up or down, he could move the turret and guns in the same directions. The right-hand arm of the controller was a twist-grip to adjust the range input of the sight, and the left arm contained a safety switch, which had to be kept pressed in to supply power to the turret. Trigger switches for the guns were operated by the gunner's index fingers. Ammunition was stored in six cans positioned below the housing, each holding 125 rounds and fed up to the guns via rollers and ammunition booster units. The cans were linked giving 375 rounds per gun. The gunner's communication, oxygen and heated suit power socket were also

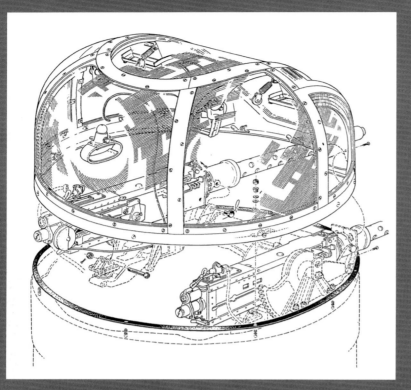

A weakness of the Fortress was its vulnerability to frontal attacks. This was addressed in the G series with the remotely controlled chin turret, which had been developed on the experimental bomber escort version of the Fortress, the YB-40. Although the programme was not a success, elements from it, in particular the chin turret, were successfully incorporated into B-17G production. The turret was controlled by the bombardier who sat directly above it in the nose of the aircraft. His controller was on the end of an arm which was stowed to the right-hand side of the aircraft when not in use. To operate it, he swung the arm to the central position and switched on the main power to the turret. By closing at least one safety switch the turret could be moved in the same direction as the controller. The unit was powered by electric motors and could rotate the two .50-calibre machine guns 86° in azimuth from the centre to each side, and 26° in elevation and 46° in depression. The guns ran in slots protected by zippered covers. Ammunition was carried internally and totalled 365 rounds per gun. The bombardier used an N-6 reflector gun sight or, in very late aircraft, a K-13 sight, suspended from a cradle at the top of the nose to aim the turret guns.

Tail gun defensive position

The redesign of the E series allowed room for a tail gunner with two .50-calibre guns to defend the rear of the Fortress. This was not a

ABOVE Diagram showing the procedure for removal of the A-1A turret dome. The machine gun mounts and charging handles are also shown. *(USAF)*

contained in the housing. The first turrets came with a low dome, which restricted the headroom for the gunner, as well as having metal side panels, which reduced visibility. Reacting to complaints from gunners, the dome went through no fewer than six design changes before a satisfactory style, known as the Ainsworth dome, with a much higher top and far less bracing bars, became standard. Another advantage was that it was constructed of a one-piece stamped metal frame, which reduced weight compared to the earlier cast-metal frames.

RIGHT Chin turret ammunition is carried internally. Up to 450 rounds per gun could be accommodated. *(USAF)*

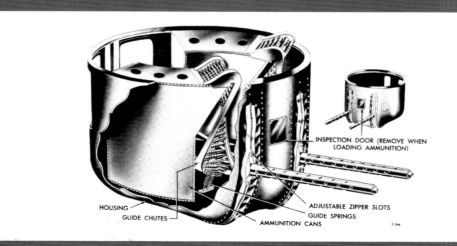

HOUSING
GUIDE CHUTES
AMMUNITION CANS
GUIDE SPRINGS
ADJUSTABLE ZIPPER SLOTS
INSPECTION DOOR (REMOVE WHEN LOADING AMMUNITION)

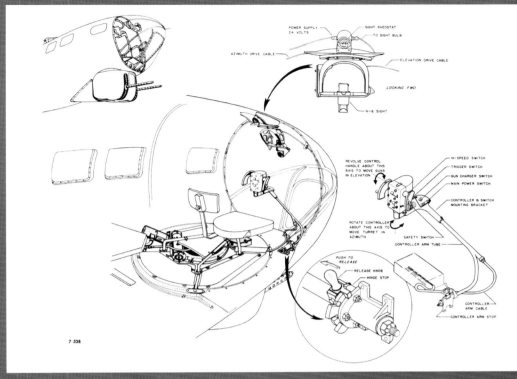

LEFT General
arrangement of
the chin turret
and controls. The
controller arm swings
to the right and fixes
to the side of the nose
frame when not in use.
(USAF)

true turret because the guns were manipulated
by the gunner and there was no power
assistance to aid him. The tail position went
through one major design change during its
use on the B-17 E, F and G series. The original
tail 'stinger' had a simple ring and bead gun
sight and a restricted range of movement
for the guns. In 1944 this was replaced by
the 'Cheyenne' tail, so-called because it was
designed by the United Airlines modification
centre at Cheyenne, Wyoming. It featured
better visibility for the gunner along with a
greater range of movement for his guns and
the installation of a more accurate reflector
gun sight. The Cheyenne tail was often
retrofitted to aircraft in the field.

FAR LEFT The tail
stinger, here shown on
The Pink Lady, was the
original tail defensive
position. *(Franck Talbot)*

LEFT The Cheyenne
tail position installed
in the later B-17G
series features more
room for the gunner
and an increased
range of movement
for the guns. It also
uses a more accurate
reflector sight. This
example is fitted to
Mary Alice.

Wing

Each wing is constructed in three parts – the tip, the outboard section containing the aileron, and the inboard section containing the flap and the engine nacelles. Attached to the firewall on each nacelle are the tubular steel engine mounts, which are interchangeable. Two spars of truss construction extend from the fuselage to the wing tip. Each spar is attached to the fuselage at an upper and lower terminal to attachment points at stations 4 and 5 by steel taper pins; the outboard wing section is similarly attached by taper pins. The wing sections can be dismantled by removing these pins. Wing ribs are numbered from 1 at the fuselage join to 37 at the wing tip.

The all-metal wing flap is a split trailing edge

ABOVE Staff Sgt A.G. Vailancourt uses a mechanical poppit riveter to repair the aluminium skin on the upper wing surface of a 92nd BG B-17 at Alconbury, Cambridgeshire, in June 1943. *(IWM D15141)*

ABOVE RIGHT The internal structure of the B-17 wing. *(USAF)*

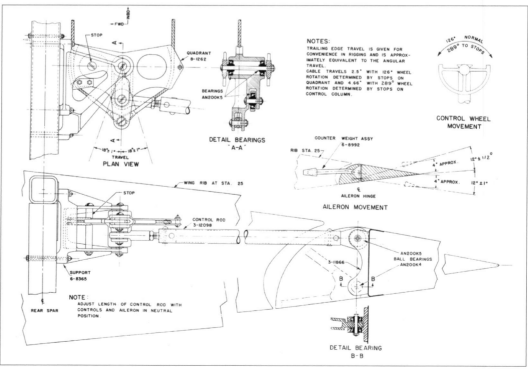

RIGHT Rigging diagram for the aileron mechanism. *(USAF)*

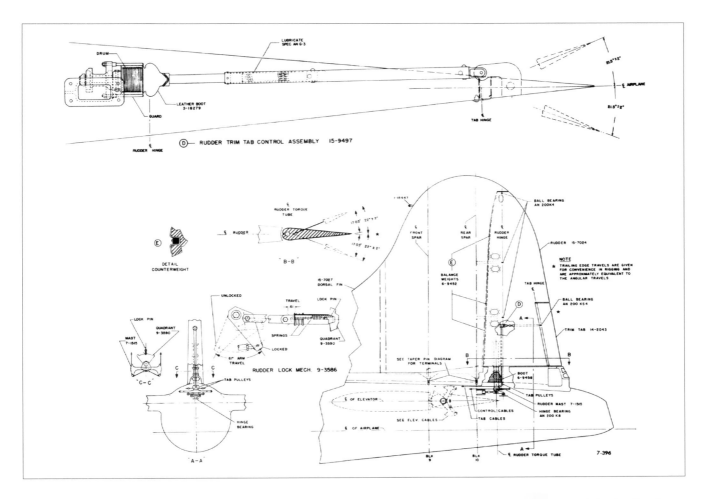

RUDDER TRIM TAB CONTROL ASSEMBLY 15-9497

design, operated by an electric motor in the left wing which drives both flaps simultaneously through a retracting screw and connecting rods. The ailerons are fabric-covered with a metal structure and dual cable operated from the cockpit, the cables controlling a quadrant and pushrod actuator. A trim tab fitted to the left-hand aileron is actuated through cables driving a drum and screw arm.

Empennage

This comprises the vertical stabiliser and rudder along with the horizontal stabilisers and elevators. The rudder and elevators are fabric-covered and have all-metal trim tabs, while the fin and stabilisers are all-metal stressed skin construction. All control surfaces are operated by a dual cable system from the two pilots' controls; trim tabs are also cable-controlled, the cables driving drums and screw actuators.

The extension of the vertical stabiliser

ABOVE Details of the rudder control mechanism. *(USAF)*

LEFT The large vertical stabiliser and rudder seen here on *The Pink Lady* **became the hallmark of the B-17E series onwards.**

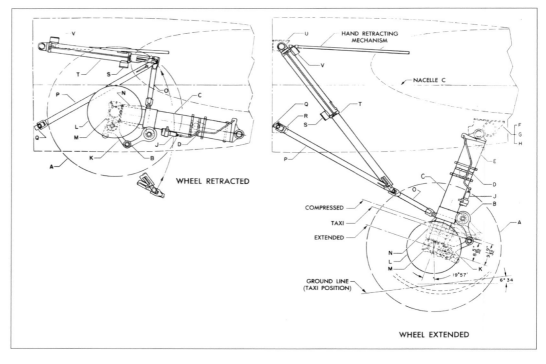

RIGHT The main wheel retracts forwards, leaving a portion of the tyre extending below the nacelle. *(USAF)*

WHEEL RETRACTED

HAND RETRACTING MECHANISM

NACELLE C

COMPRESSED
TAXI
EXTENDED

GROUND LINE (TAXI POSITION)

WHEEL EXTENDED

BELOW A Vega employee works on the starboard main wheel. *(Lockheed)*

running along the top of the fuselage is known as the dorsal fin and it is attached to the vertical stabiliser at fuselage station 8.

The horizontal stabilisers are constructed in a similar manner to the wings using two truss-type spars, and attached to the fuselage using taper pins. The left and right stabilisers are interchangeable.

Landing gear

The landing gear comprises two main units housed in the inboard nacelles and a tail unit at fuselage station 7. All three units are retractable.

The main undercarriage legs consist of an oil air shock strut or oleo, torsion links, a drag strut assembly and a retraction screw. Retraction is accomplished electrically by a 24-volt motor at the top of the nacelle driving through a 40 to 1 reduction gearbox. Full extension and retraction of either leg should take no longer than 45 seconds and is controlled by limit switches. An emergency hand crank system for each leg is installed, consisting of a series of drive shafts and right-angled gearboxes, which allow the leg to be lowered or raised manually from a crank installed in the rear face of station 4. Each leg is fitted with a 56in-diameter tyre and duplex expander tube brakes. The shock strut is inflated with compressed air from a source developing at least 800psi.

The tail unit consists of a retracting screw jack, a yoke, the oleo strut, treadle frame and the knuckle assembly carrying the axle. Retraction is achieved electrically using the same type of motor as for the main gear, and emergency operation is via an extension shaft and hand crank socket aft of the wheel. There is a switch to prevent the

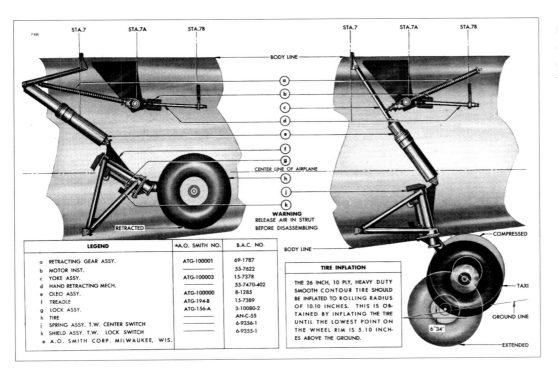

STA.7 · STA.7A · STA.7B · STA.7 · STA.7A · STA.7B

7 230

BODY LINE

(a) (b) (c) (d) (e) (f) (g)

CENTER LINE OF AIRPLANE

(h) (i) (k)

RETRACTED

WARNING
RELEASE AIR IN STRUT
BEFORE DISASSEMBLING

COMPRESSED

BODY LINE

TAXI

GROUND LINE

6°34'

EXTENDED

LEGEND		*A.O. SMITH NO.	B.A.C. NO.
a	RETRACTING GEAR ASSY.	ATG-100001	69-1787
b	MOTOR INST.		55-7622
c	YOKE ASSY.	ATG-100003	15-7378
d	HAND RETRACTING MECH.		55-7470-402
e	OLEO ASSY.	ATG-100000	8-1285
f	TREADLE	ATG-194-B	15-7389
g	LOCK ASSY.	ATG-156-A	3-10080-2
h	TIRE		AN-C-55
i	SPRING ASSY. T.W. CENTER SWITCH		6-9356-1
k	SHIELD ASSY. T.W. LOCK SWITCH		6-9355-1
*	A. O. SMITH CORP. MILWAUKEE, WIS.		

TIRE INFLATION

THE 26 INCH, 10 PLY, HEAVY DUTY SMOOTH CONTOUR TIRE SHOULD BE INFLATED TO ROLLING RADIUS OF 10.10 INCHES. THIS IS OBTAINED BY INFLATING THE TIRE UNTIL THE LOWEST POINT ON THE WHEEL RIM IS 5.10 INCHES ABOVE THE GROUND.

LEFT The tail wheel retracts backwards and fully into the fuselage. *(USAF)*

wheel from retracting unless it is centred, and another to indicate to the pilots that the locking pin is in place. A cable is connected to the floor lever in the cockpit which, when lifted, withdraws the locking pin allowing the tail wheel to rotate through 360° for taxiing. If the lock is inadvertently left in when the aircraft taxies, a special shear bolt is designed to break, preventing damage to the locking pin. The tail wheel tyre is a 26in-diameter smooth contour.

Fuel system

The fuel system comprises four independent supplies, each feeding one engine. Each of the outboard engines is fed

LEFT Diagram showing the complete fuel system and controls. The wing tanks labelled 'T' and 'U' were commonly known as Tokyo tanks. *(USAF)*

A No 1 & 4 engine tank
B No 2 & 3 engine tank
C feeder tank
D booster pump
E fuel strainer
F engine fuel pump
G fuel pressure transmitter
H oil dilution valve
I carburettor
J fuel transfer select valve
K transfer pump
L bomb-bay fuel tank
M primer
N restriction fitting
O tank vents – underside of wing
P fuel shut-off valve
Q tank filler neck
R oil drain cock assembly
S drain cock
T outboard wing tanks Nos 1–6
U inboard wing tanks Nos 7–9
V valve – transfer series A
W control – remote manual
X sylphon
Y tank drain valve
Z fuel quantity gauge

NOTES
ALL SELF-SEALING HOSE SHALL BE IN ACCORDANCE WITH AIR CORPS SPEC. 26587 AND SHALL SATISFACTORILY RESIST DETERIORATION FROM AROMATIC FUEL WITH A COMPOSITION OF 60% 100 OCTANE GASOLINE, 20% TOLUENE, 15% XYLENE AND 5% BENZENE. TUBING COLOR IDENTIFICATION - RED

FUEL SHUT OFF VALVE
OPEN CLOSED
ON FUEL BOOST PUMP OFF

ON CENTRAL CONTROL PANEL

OFF
ON 1 ON 2
ENGINE PRIMER
ON 4 ON 3

ON COPILOT'S RIGHT SIDE WALL

* B.I.G.—BRITISH IMPERIAL GALLONS

SWITCH
FUEL TRANSFER PUMP
L.H. TANKS — R.H. TANKS TO — OFF — TO
R.H. TANKS — L.H. TANKS
LOCATED BELOW CONTROL CABIN REAR DOOR

TRANSFER VALVES · LOOKING AFT

ABOVE One of the two inboard fuel tanks removed. This tank is shown as 'B' in the fuel system diagram on page 89 and held 213 US gal.

from a 425gal tank, while the inboard engines receive their supply from one 213gal engine tank and a 212gal feeder tank. Later B-17s had nine additional fuel cells installed between the outboard wing ribs. These so-called 'Tokyo' tanks increase fuel capacity by a further 1,080gal making the total fuel load 2,780gal. For special long-range missions it is possible to carry two extra tanks in the bomb-bay of 410gal each, taking the absolute maximum fuel load to 3,600gal, which equates to a theoretical endurance of over 22 hours without a bomb load. Fuel quantity is measured on a gauge on the co-pilot's instrument panel and indicates the quantities in each of the four engine tanks and feeder tanks. The Tokyo tanks, being arranged to feed into the main tanks, have no fuel gauges. Also on the co-pilot's panel are the fuel pressure gauges. On the forward panel of the centre pedestal there are four toggle switches to operate the fuel booster pumps, one for each engine,

which provide fuel for starting and act as back-up for the mechanical fuel pumps for take-off and landing. Alongside these switches are four more which operate the emergency fuel shut-off valves. These will normally always be open.

Fuel can be transferred in flight from one tank to another, for the purpose of keeping the fuel load balanced. This is accomplished by a fuel transfer pump situated in the bomb-bay. The switches and valves are positioned behind the pilots on the forward face of station 4. Fuel can only be transferred across the aircraft centre line, if it is desired to transfer between two tanks in one wing, then two transfer steps must be made.

Fuel consumption at normal cruise speed of 150mph is approximately 150 to160gals/hr.

Oil system

Each engine is supplied from its own independent oil system comprising a 37-US gal tank behind the engine in the nacelle, an oil cooler installed in the leading edge of the wing and two engine-driven pumps; the pressure pump which draws oil from the tank and then forces it under pressure to the moving parts of the engine; and the scavenge pump which draws off the oil that has drained into the sump and returns it via the oil cooler back to the supply tank. The tank is equipped with a metal cylinder or 'hopper', which extends between the inlet and outlet openings. When the engine is first started, warm oil from the engine enters the hopper and flows immediately to the outlet and back to the engine, reducing engine warming-up time. Some cold oil is permitted to enter the hopper through openings in the top and bottom of the cylinder.

On the co-pilot's instrument panel there are oil pressure and oil temperature gauges for monitoring the oil system.

The oil cooler regulates the amount of cooling of the oil automatically by a temperature regulator valve, which responds to oil pressure and temperature and adjusts shutters at the rear of the cooler to regulate the airflow and hence the amount of cooling. When the oil is cold such as at start-up, the regulator directs the oil around the outside jacket of the cooler, or muff; as the oil warms up, it is sent through the central, cooling section, the core.

RIGHT The oil cooler in this photograph has been disconnected and is about to be removed from the wing. The unit at the top is the temperature regulator valve which controls the airflow through the unit.

Propellers

Propellers are of the three-blade Hamilton Standard, Hydromatic, constant-speed, full-feathering type. The blade angle is controlled by a governor device, which maintains the speed of the propeller by adjusting the pitch of the blades through regulation of the oil flow to a piston contained in the dome of the propeller assembly. The governor is linked by a cable to the propeller pitch control lever in the cockpit, so that when the pilot adjusts it to a new setting, the governor will readjust the oil flow to move the piston in the dome, which re-sets the blades to a new pitch angle.

Propeller pitch range is: low blade angle 20°, high blade angle 80°, and the fully feathered angle is 88°.

If an engine needs to be shut down in flight, the propeller blades must be rotated edgeways on to the airflow, or feathered. This is achieved by the use of a feathering pump situated behind the engine on the firewall. When operated by a push-button in the cockpit, the pump draws oil from the main engine oil supply line and sends it at a pressure of around 400psi to the piston in the dome. This forces the piston to drive the blades to the feathered position and stop them rotating. Failure of the feathering system can be serious. If it is not possible to stop the propeller rotating, then excessive drag is created leading to the possible detachment of the propeller from the engine.

Hydraulic system

The services supplied by the hydraulic system are the main wheel brakes and the engine cowl flaps. The hydraulic panel is located on the sidewall behind the co-pilot's seat and consists of the hydraulic pump, a relief valve, shut-off valve, check valve, filter and pressure regulating switch.

The hydraulic pump is operated from the aircraft's 24-volt electrical supply. It supplies hydraulic fluid at a rate of 1gal/min to the hydraulic accumulator, located on the forward face of station 4, which acts as a storage unit to supply pressure to the system. The system pressure is maintained between 600 and 800psi by the pressure regulating switch. Next to the

LEFT Inside the propeller hub – the three bevel gears rotate the propeller blades to change the pitch angle.

accumulator on the bulkhead is the supply tank which holds 1¾gal of fluid.

The cowl flaps are actuated on each engine by a hydraulic cylinder operating through rods and bellcranks. Situated forward of the carburettor at the top of the nacelle, the cylinder has hydraulic fluid admitted to either end by the operation of the control valves in the cockpit. The valve has 'open', 'closed' and 'locked' positions.

The brakes are supplied via a de-boost valve at the rear of each landing gear oleo strut. This valve reduces the hydraulic system pressure down from around 800psi to 200psi which operates the brake expander tubes.

For emergency operation, if the hydraulic pump fails, a hand pump is located to the right of the co-pilot's seat. A hydraulic pressure gauge is located in a console in the centre of the cockpit ceiling and a low-pressure warning light is adjacent to it.

BELOW Diagrams showing the hydraulic panel and the supply tank situated at the rear of the cockpit. The hydraulic pump supplies pressure between 600 and 800psi to the cowl flaps and wheel brakes. *(USAF)*

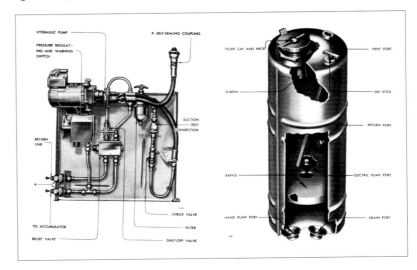

RIGHT The later cabin air heating system involves ducting hot air from the two inboard engines to heat exchangers in the wing, from where it is distributed throughout the aircraft. *(USAF)*

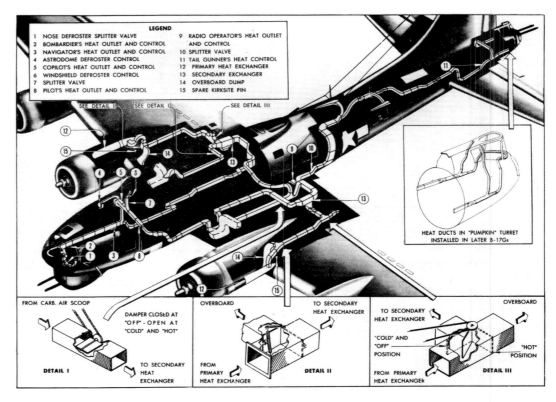

LEGEND

1 NOSE DEFROSTER SPLITTER VALVE
2 BOMBARDIER'S HEAT OUTLET AND CONTROL
3 NAVIGATOR'S HEAT OUTLET AND CONTROL
4 ASTRODOME DEFROSTER CONTROL
5 COPILOT'S HEAT OUTLET AND CONTROL
6 WINDSHIELD DEFROSTER CONTROL
7 SPLITTER VALVE
8 PILOT'S HEAT OUTLET AND CONTROL
9 RADIO OPERATOR'S HEAT OUTLET AND CONTROL
10 SPLITTER VALVE
11 TAIL GUNNER'S HEAT CONTROL
12 PRIMARY HEAT EXCHANGER
13 SECONDARY EXCHANGER
14 OVERBOARD DUMP
15 SPARE KIRKSITE PIN

HEAT DUCTS IN "PUMPKIN" TURRET INSTALLED IN LATER B-17Gs

DETAIL I — FROM CARB. AIR SCOOP — DAMPER CLOSED AT "OFF" - OPEN AT "COLD" AND "HOT" — TO SECONDARY HEAT EXCHANGER

DETAIL II — OVERBOARD — TO SECONDARY HEAT EXCHANGER — FROM PRIMARY HEAT EXCHANGER

DETAIL III — TO SECONDARY HEAT EXCHANGER — OVERBOARD — "COLD" AND "OFF" POSITION — "HOT" POSITION — FROM PRIMARY HEAT EXCHANGER

Heating and ventilation

The aircraft is provided with a heating system which conducts warm air throughout the aircraft. Hot air surrounding the exhausts of the two inboard engines is ducted to heat exchangers in the wings; from these, the warmed air is directed around the fuselage. If heating is not required, the hot air from the exhaust is dumped overboard through slots in the top of the wing. Controls to regulate the heat and airflow are operated via levers in the radio room, which actuate cables that position valves in the duct system. Warm air is directed to all parts of the aircraft to heat compartments and defrost windows, from the bombardier's flat panel in the nose to the tail gunner's windows.

Earlier aircraft had a different system which was largely ineffective aft of the radio room. Conditions in the rear part of the aircraft were usually very cold due to the open waist windows. The system relied on pumping ethylene glycol fluid from a tank in the nacelle of number two engine through heaters in the exhaust system of the same engine.

De-icer and propeller anti-icer systems

The wing, horizontal stabiliser and fin leading edges are fitted with rubber de-icer shoes, which are alternately inflated and deflated in order to break up any ice that builds up on these surfaces. They are operated by vacuum pumps installed on numbers two and three engines. The shoes are inflated by the exhaust from both pumps, and then deflated by suction from one pump. The process is controlled by a motor-driven distributor valve, which alternately redirects the suction and pressure to the shoe distribution lines. The pilot can select which of the two vacuum pumps is used to deflate the shoes by a control valve to the left of his seat. Vacuum is also used to operate gyro-powered flight instruments and the camera equipment.

To prevent the formation of ice on the propellers an anti-icing system is installed. Consisting of a tank containing an alcohol-based fluid and two pumps located under the radio compartment floor, the fluid is pumped to slinger rings at the base of the propeller blades. The pump speed is controlled by rheostats on the floor panel to the left of the pilot's seat.

Autopilot

A Type C-1 autopilot is installed, which maintains the aircraft in straight and level flight and allows it to be manoeuvred using fingertip control by the pilots and the bombardier. The system consists of the pilots' control box mounted in the front of the central control pedestal, two gyros, one under the pilots' compartment floor, the other in the bomb sight stabiliser panel, an amplifier, a DC to AC inverter, the pilot's PDI instrument and three servo motors to actuate the ailerons, elevators and rudder. With the autopilot switched on, deviation of flight about any of the three axes is sensed by movements within the gyros. Signals are sent to the amplifier, which boosts these signals and sends them to the appropriate servos to compensate for the deviation. Once the aircraft has returned to level flight, the gyros return to their stabilised position and the compensating signals stop. The Pilot Directional Indicator informs the pilot that the autopilot and aircraft are correctly trimmed. On the bomb-run it indicates any course corrections required by the bombardier.

In a precursor to modern aircraft control systems with their side sticks (for example those used in Airbus airliners), late B-17Gs are fitted with formation sticks. These allow the pilots to control the aircraft through a fighter-style column grip incorporating an armrest, and the whole unit is mounted to the cockpit floor. The sticks are connected into the autopilot control system.

Electrical system

The primary source of electrical power is a 24-volt DC system. Power is derived from three 24-volt, 34-amp/hr batteries in the leading edges of the wing, and four engine-driven, 200-amp generators attached to the accessory casing of each engine. Additionally, there is a 400Hz, 115-volt AC system supplied from one of two motor-driven inverters, which supply AC power to the turbo supercharger control system in later aircraft.

The following services are supplied by the DC electrical system:

LEFT **The formation stick was a late development, allowing pilots to fly the aircraft using one hand. It was operated through the autopilot system.**

- Hydraulic pump
- Landing gear
- Flaps
- Bomb doors and bomb-release system
- Fuel shut-off valves
- Fuel booster pumps
- Fuel transfer pump
- Fuel quantity gauge
- Oil dilution valves
- Inertia starters
- Propeller feathering pumps
- Propeller anti-icing pumps
- De-icer distributor valve
- Carburettor filter valves
- Engine temperature instruments
- Pitot heater
- Landing lights
- Navigation lights
- Interior lights
- Radio and navigation equipment
- Gun turrets

For ground use, a portable auxiliary generator is carried in some aircraft. All aircraft have an external power plug for connecting a ground power source. The plug is located just aft of the crew entry hatch in the nose.

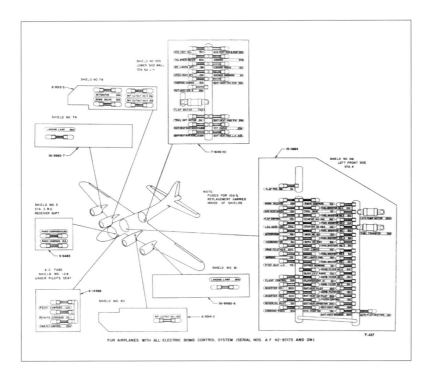

Electrical circuit distribution and control is via a fuse panel on the forward bulkhead of station 4, and either a fuse panel or a circuit breaker panel at station 6, and an AC fuse shield under the pilot's seat. Control switches for the batteries, generators and inverters are on a box to the pilot's left. The fuses for the landing lights, battery solenoid switches, and for the IFF radio detonator are located either in the nacelles or the battery compartments and so are not accessible in flight. All circuits, except for the starter motors, landing gear retraction motors, bomb door motor and the propeller feathering motors, are protected by fuses. Reliable operation of these circuits was considered more important than circuit protection.

Emergency equipment and procedures

The emergency equipment carried on board a Fortress during the Second World War comprised the following systems and items:

Hand cranks
Hand cranks for the emergency operation of the main landing gear and tail wheel, wing flaps and bomb-bay doors are attached to the forward face of the aft bulkhead of the radio room. The main wheels are hand cranked from positions inside the bomb-bay, on either side of the entry door to the pilots' compartment; the tail wheel is raised or lowered from the tail wheel compartment; the flaps from a recess under the radio room floor; and the bomb-bay doors from a socket located at the forward end of the bomb-bay catwalk.

Bomb release
The bomb load could be salvoed unarmed by operating one of three salvo switches, on the bombardier's panel, one above the instrument panel in the cockpit or one in the bomb-bay. Earlier aircraft had a cable release system.

Fire extinguishers
Three portable extinguishers are carried: in the navigator's compartment, in the cockpit and in the radio room. The engine fire extinguishing system comprises two carbon dioxide fire bottles installed under the pilots' floor. These can be routed to any of the four engines via a control panel on the co-pilot's side of the cockpit.

Signalling
In the event of a communication failure, the pilot could send a warning signal to the crew by three alarm bells – one each under the navigator's and radio operator's tables and the third in the tail compartment.

First-aid kits
Three first-aid kits were carried – one in the navigator's compartment, one in the cockpit behind the co-pilot's seat and another in the waist area forward of the ball turret.

Bailing out
The crew could leave the aircraft from four parachute exits – the forward entry hatch, through the bomb-bay, the entry door in the waist and the tail gunner's hatch under the tail.

The door and hatches each had an emergency jettison handle which, when pulled, withdrew the hinge pins allowing the hatch to be released.

SECTION III
EMERGENCY INSTRUCTIONS

LEFT Emergency instructions include information on the use of the hand cranks to lower the landing gear. *(USAF)*

Figure 30—Hand Cranks Stowed—Right Side

1. HAND CRANKS.

Cranks for manual operation of landing gear, wing flaps, and bomb bay doors, and for hand starting of engines, are stowed on the aft bulkhead of the radio compartment. Crank extensions for use when operating engine starters, bomb doors, and wing flaps are stowed adjacent to the cranks.

2. EMERGENCY OPERATION OF LANDING GEAR.

Each main landing gear is operated through hand crank connections to the left and right of the door in the bomb bay's forward bulkhead.

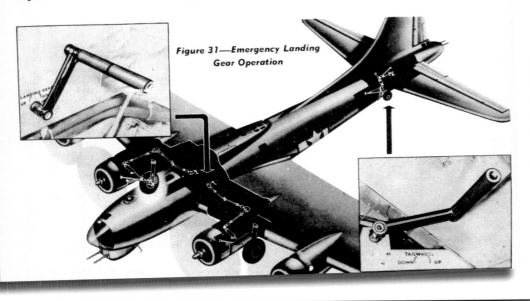

Figure 31—Emergency Landing Gear Operation

BELOW If the aircraft needed to perform a wheels-up landing it was important to drop the ball turret first to prevent it breaking the back of the fuselage. This placard, which was placed in the waist area of the fuselage, gives instructions on how to remove the turret. *(via Dave Littleton)*

INSTRUCTIONS FOR EMERGENCY DROPPING OF BALL TURRET IN FLIGHT

TO DROP TURRET :

1.   REMOVE COMPUTING SIGHT, IF TIME PERMITS.

2.   DISCONNECT ELECTRICAL AND OXYGEN LINES, IF TIME PERMITS.

3.   REMOVE FOUR NUTS AND AZIMUTH GEAR CASE.

4.   BREAK OFF OR REMOVE ALL SAFETY RETAINING HOOKS OVER RING GEAR.

5.   REMOVE TWELVE YOKE CONNECTING BOLTS. IF TURRET HANGS ON FIRE CUT OFF CAM, A SWIFT KICK DOWN WILL DISLODGE IT.

FOR ADDITIONAL DETAILS, CONSULT T.O. No. OI-20EG-I

13. HOW TO DITCH THE B-17G:

a. Jettison bombs, ammunition, and loose equipment and secure anything that might cause injury. Close bomb doors and lower hatches. If time is too short to release bombs and depth charges, place them on SAFE. Keep enough fuel to make a power landing.

b. Navigator gives radio operator position, course and speed. Latter tunes liaison set to MFDF and SOS's position and call sign continuously. R/O also turns IFF to distress, remains on intercomm., and clamps down key on order to "take to ditching post".

c. Here's how to tell wind direction and speed: (a) waves in open sea move downwind: (b) spray direction indicates wind direction; (c) wind lanes (series of lines or alternate strips of light and shade) also show direction; (d) approach on waves should be made into wind at right angles to them; (e) approach on swells should be made along top, parallel to swell and may be executed in winds not over 10 MPH.

HOW TO DETERMINE WIND SPEED

A few white crests	10 to 20 MPH.
Many white crests	20 to 30 MPH.
Foam streaks on water	30 to 40 MPH.
Spray from crests	40 to 50 MPH.

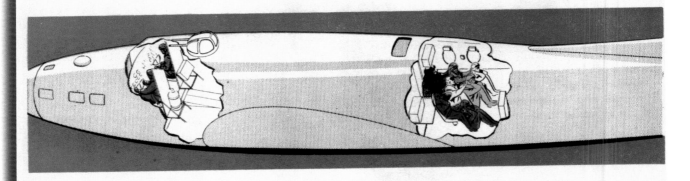

d. These positions should best enable crew members to withstand the impact of crash landings on either land or water. On water two impacts will be felt, the first a mild jolt when the tail strikes, the second a severe shock when the nose strikes the water. Positions should be maintained until the aircraft comes to rest. Study them carefully.

Emergency equipment for use in the dinghy should be carried to crash positions. Any equipment carried free must be held securely during ditching to prevent injury.

Parachute pads, seat cushions, etc., should be used to protect the face, head, and back.

ABOVE An outline of ditching procedures. All crew members other than the pilots gather in the radio compartment. *(USAF)*

96

BOEING B-17 FLYING FORTRESS MANUAL

Parachutes

Crew members could wear either a back or seat parachute; however, in most crew positions these were too bulky to allow freedom of movement and in the case of the ball turret it was impossible to wear one. Each crew member stowed a personal parachute and dingy pack close to their crew station, ready to clip on to a quick attachable chute harness if the order was given to abandon the aircraft.

Crash-landing

In the event of a crash-landing, the crew were advised that those in seats should fasten their seat belts prior to impact. Gunners braced their backs against the radio room bulkhead. The crew could vacate the aircraft via the pilots' sliding windows and the radio room hatch.

Emergency release of the ball turret

If the landing gear could not be lowered and a wheels-up landing was inevitable, then the crew were advised to drop the ball turret to prevent serious structural damage to the fuselage. Aircraft were frequently repaired and returned to service after a wheels-up landing providing the turret had been dropped. Tools and instructions were carried inside the aircraft.

Ditching

For a water landing, all crew other than the pilots gathered in the radio room. Impact with the water was often severe and the aircraft usually sank rapidly, so the crew had to act quickly. Two five-man dinghies were stowed above the bomb-bay and could be released either from inside the radio room or from outside the aircraft. A portable emergency radio transmitter, Type SCR-578-A, was carried, which was enclosed in a waterproof buoyant case and transmitted on the distress frequency of 500kHz.

Other emergency equipment

A crash axe was fitted to the seat frame of the pilot's seat. At the rear of the cockpit, behind the co-pilot, a Very pistol mount was installed to allow for emergency signalling. Cartridges for the pistol were contained in a bag in the cockpit.

e. KNOW YOUR DITCHING DUTIES! PRACTICE THEM! DRILL IS IMPORTANT!

ABOVE Life-raft drill and evacuation of the aircraft after ditching. *(USAF)*

BELOW Lt C.J. Kelly (left) and Cpl A.M. Gargano of the 379th BG, 525th BS, inspect a rubber dinghy to be carried in B-17F, 42-29511, *The Iron Maiden*, pictured at Kimbolton, Cambridgeshire, in June 1943. *The Iron Maiden* was shot down on the Schweinfurt raid on 14 October 1943 (*see page 49*). All ten crew became POWs. *(IWM D15114)*

BLADE DWG : 6477A-0
BLADE MFG : J-1365
● LOW ANGLE : 20
● HIGH ANGLE : 88

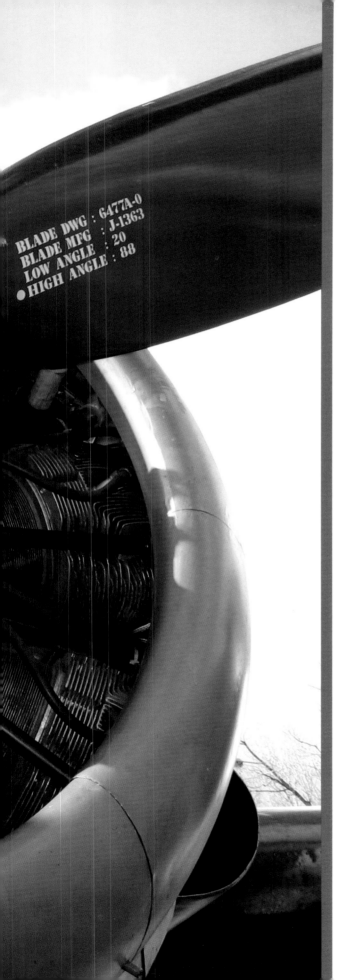

BLADE DWG : 6477A-0
BLADE MFG : J-1363
LOW ANGLE : 20
● HIGH ANGLE : 88

Chapter Five

The Wright Cyclone R-1820-97 Engine

The Wright Cyclone radial engine was a reliable power plant with a good power-to-weight ratio, which served the B-17 well in its high altitude missions during the Second World War. The engines are still powering today's airworthy Fortresses in friendlier skies more than 80 years after they were first designed.

OPPOSITE The nine-cylinder Cyclone was a robust and reliable power plant for the Fortress. *(Franck Talbot)*

History and development

The development of the Cyclone goes back to the Twenties when the Wright Aeronautical Corporation were designing a successor to the J-5 Whirlwind, nine-cylinder radial engine of the type which powered Charles Lindbergh's Ryan NYP monoplane, *Spirit of St Louis,* across the Atlantic in 1927. The first of these replacement designs, also a single-row, nine-cylinder engine, and the first to be named Cyclone, was the R-1750. The type designation 'R' denotes a radial engine with a capacity of 1,750cu in, which was tested at 500bhp in 1927. An enlarged version of the same engine, the R-1820, was developed with a bore of 6.125in and a stroke of 6.875in, giving a capacity of 1,823cu in, and by 1939, thanks to further development (chiefly in the design of cylinder cooling fins), was reaching around 1,000bhp.

As installed in the B-17, when fitted with

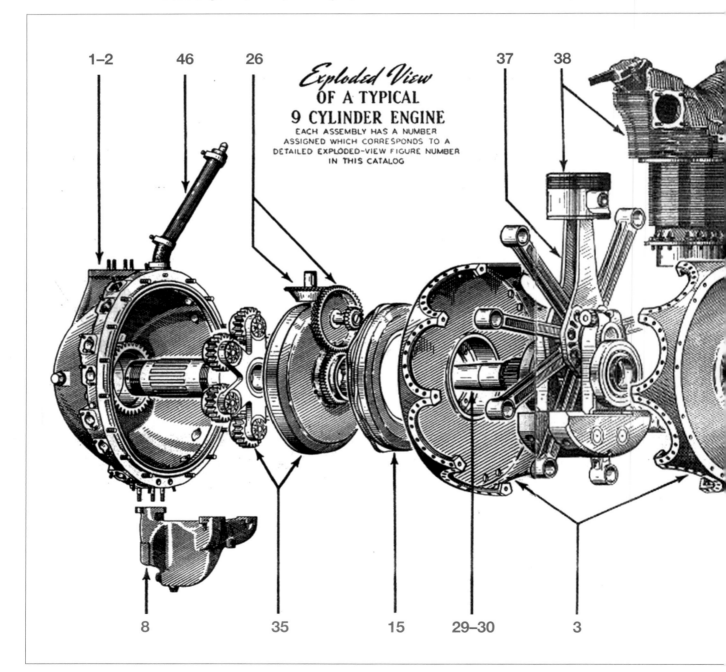

1–2   46   26   37   38

Exploded View
OF A TYPICAL
9 CYLINDER ENGINE
EACH ASSEMBLY HAS A NUMBER
ASSIGNED WHICH CORRESPONDS TO A
DETAILED EXPLODED-VIEW FIGURE NUMBER
IN THIS CATALOG

8   35   15   29–30   3

a turbo supercharger to supplement the mechanically driven internal supercharger, the R-1820 is rated at 1,200bhp at 2,500rpm for take-off. The compression ratio is 6.7:1 and the supercharger ratio is 7:1. The dry weight of the engine is 1,308lb.

The two versions of the R-1820 were fitted to the majority of B-17s – the -65 and the -97 were installed into the E, F and G series, differing only in the addition of an external

oil scavenge line fitted between the oil sump strainer and the oil pump on the latter engine.

BELOW Exploded view of a Wright Cyclone from the USAF engine parts manual. The numbers referred to in the following photograph captions on pages 102–6 correspond to those in this exploded view. (All photographs Graeme Douglas unless credited otherwise)

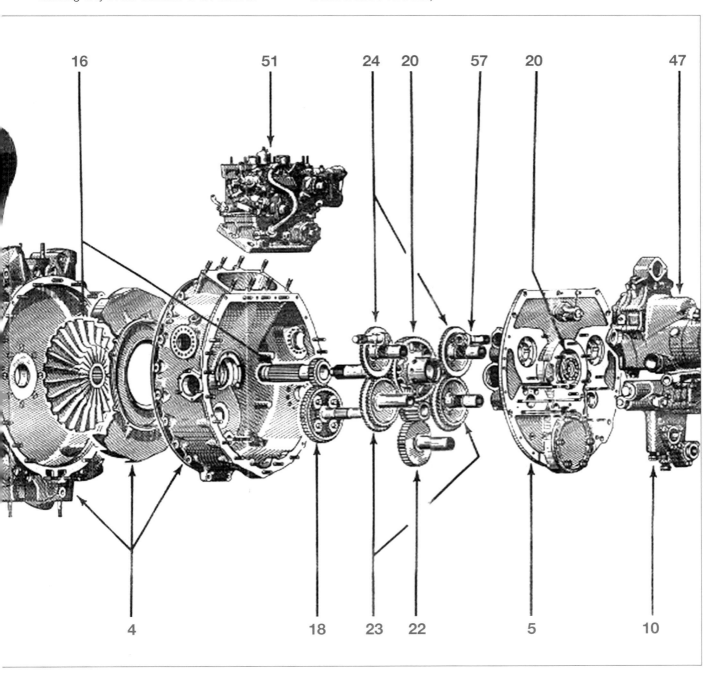

The engine

The crankcase and crankshaft

The crankcase consists of six sections: the crankcase front cover housing the reduction gear, the front and rear main sections, the supercharger front and rear housings and the supercharger rear cover. The two main sections, constructed from alloy steel forgings, house the single-throw crankshaft which is built up in two parts of machined alloy steel forgings. The crank webs contain dynamic damper counterweights, which swing out as the engine turns to reduce vibration. The crankshaft is supported at the front and rear by roller bearings and contains a single crankpin. The plain, one-piece, steel-backed, big end bearing of the master connecting rod is pressed into the big end bore. The master rod is machined from a steel forging and is of H-section, located inside number one cylinder; and around the master rod, eight equally spaced articulated rods pivot on knuckle pins and serve the other cylinders. The knuckle pin

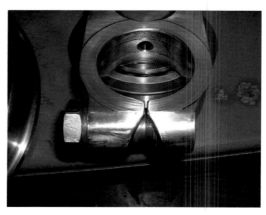

FAR LEFT The rear web of the crankshaft, complete with a main bearing and the dynamic counterweight at the bottom (29–30).

LEFT The rear main bearing resting on one of the crankcase halves (3).

FAR LEFT A big end bearing which has been re-metalled and machined to size, shown in front of the big end bore into which it locates (37).

LEFT The master rod, which operates in number one cylinder with the big end bearing located, surrounded by the eight articulated rods (37).

and piston pin ends of the rods have split-type bronze bushings pressed into them. The nine cylinders are attached by studs and nuts to the circumference of the crankcase main sections.

Cylinders

The cylinders consist of cast aluminium alloy heads shrunk on to forged steel barrels. The cylinder bores are nitride treated to increase their life. Air deflector baffle plates between the cylinders ensure that cooling air is guided between the fins. The bolting flange has 20 holes to allow each cylinder to be attached to the crankcase. Cylinders are numbered, starting with number one at the top of the engine and then counted in a clockwise direction when viewed from the rear (anti-

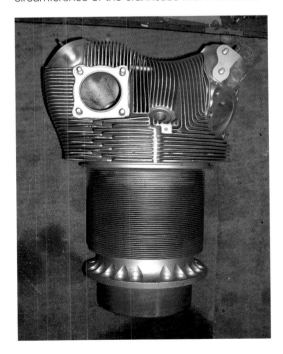

FAR LEFT A cylinder viewed from the rear, with the exhaust port on the left (38).

LEFT Close-up of the inlet rocker box assembly (38).

RIGHT The inlet port and valve (38).

valves. The valves are machined from forgings of heat resistant steel, the exhaust valve is sodium filled to aid cooling and both valves are fitted with three interchangeable concentric springs. Valve operation is accomplished by a cam ring located in the front crankcase; this is a hardened steel ring with two sets of lobes operating pushrods which are enclosed in tubes at the front of the cylinders. The pushrods operate adjustable rocker arms in the rocker boxes.

Pistons

The pistons are machined from heat-treated aluminium forgings and are of dome-headed design. They are fitted with six rings – the top three are wedge-type compression rings; the fourth and fifth act as oil control rings (their grooves having holes drilled in them to control oil flow); and the sixth ring, fitted below the piston pin, is an inverted oil control ring designed to spread a film of oil over the piston and bore. The piston pins are of alloy steel with a hardened surface; they are of tubular construction and are fully floating in the connecting rod bushings and in the pistons.

propeller) end of the engine. Two spark plug bushes at a 90° angle are installed at the top of the cylinder. Cast into the cylinder heads, two rocker boxes house the inlet and exhaust

RIGHT The pistons are fitted with six rings; the bottom ring is an inverted oil control ring (38).

BELOW LEFT The reduction gear cover with the prop shaft at the top (1–2). The apertures around the base are for the pushrod tubes (46).

FAR RIGHT View inside the reduction gear cover (1–2) showing the 20 planetary gears which drive the propeller shaft (35). At the bottom right is the gear drive for the propeller governor (26).

Reduction gear cover

The front crankcase section is made of magnesium alloy and forms a housing for the propeller reduction gears. The reduction ratio of 0.5625:1 is achieved by a single gear on the end of the crankshaft meshing with 20 planetary gears mounted on a carrier ring bolted to the propeller shaft flange. The planetary gears rotate around the rim of a stationary reduction gear causing the propeller shaft to rotate in the

ABOVE The cam gear is shrunk on to the splines of the crank shaft. In order to remove it the assembly must be heated in oil to 190–204°C for 30 minutes and then a force of 5 tons applied to break the joint (35).

ABOVE View of the cam ring (15) and reduction gears (26) on top of the front crankcase half (3). Note there are four lobes each for inlet and exhaust. The bevelled gear drives the prop governor (26).

same direction as the crankshaft, but at exactly 9/16th of its speed. The reduction gear cover houses a thrust bearing shrunk into a steel retainer ring at the front of the cover, which carries the end loads of the crankshaft. Also contained inside this cover are the propeller governor drive and the cam ring, which is driven at 1/8th of crankshaft speed by intermediate and reduction gears.

Supercharger housing

The front and rear of the supercharger housing are machined from magnesium alloy castings. The front part contains nine lugs for the engine mountings, along with nine bosses for each of the intake pipe tubes. The forward web of

the supercharger and the diffuser chamber are contained in the forward section of this housing. The rear housing forms the entrance to the impeller and is the location for the accessory case drive gears. The supercharger housing cover is machined from magnesium alloy and takes the form of a reinforced flat plate screwed to the back of the rear housing. It provides the mounting pads for the engine accessories, namely the two magnetos, generator, starter, oil pump, fuel pump, tachometer drive and vacuum and glycol pumps when fitted.

Supercharger

This is of the centrifugal type, consisting of an 11in-diameter impeller or blower, the diffuser

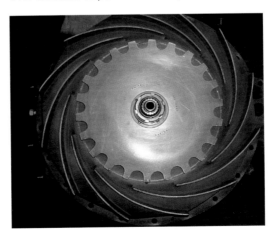

FAR LEFT The rear supercharger housing showing the gear train for the supercharger drive (4) (18) (20).

LEFT View of the rear of the supercharger impellor; the curved diffuser vanes guide the compressed mixture to the inlet tubes (4) (16).

RIGHT Inside the rear cover (5) showing the gears that drive the accessories at the back of the engine (20) (22) (23) (24).

chamber, the distribution passage and the inlet pipes. The blades of the impeller rotate in the diffuser chamber, forcing the air/fuel mixture to the outside of the chamber. Diffuser vanes guide the compressed mixture into the distribution passage from where the nine inlet tubes lead to each cylinder. The supercharger gear train is protected from overload by the incorporation of friction clutches in the intermediate gears.

Oil sump

Between cylinders five and six, a magnesium alloy cast sump is installed which, when the

BELOW The down-draught carburettor mounted to the rear of the engine (51).

engine is stationary, holds approximately 1gal of oil. The sump contains a filter screen and a magnetic drain plug to trap foreign and metallic particles in the oil.

Accessory drive mechanism

All the rear-mounted engine accessories and the supercharger are driven by a forged steel extension shaft from the rear of the crankshaft through an accessory drive gear attached to the drive shaft. At the rear end of the shaft, a three-jawed starter dog permits operation of the electric starter motor. An intermediate gear train operating through friction clutches drives the supercharger impeller at seven times engine speed. Drive to all accessories is via spur gears mounted in the rear supercharger cover, supported by bushings fixed in the rear cover.

Carburettor

A Bendix Stromberg, down-draught injection carburettor, Type PD12H2 or PD12112, is fitted to the top of the rear supercharger housing. The engine-driven fuel pump delivers fuel to the regulator unit of the carburettor at about 14psi; the carburettor controls fuel to the engine according to the amount of air flowing through it. Compensation for altitude is made by a pressure bellows unit which reduces the fuel flow to the jets as the air pressure reduces with altitude. The pilot has control of the mixture setting, allowing him to select 'Auto Rich', 'Auto Lean' or 'Full Rich', dependent on engine power settings. The carburettor discharge nozzle sprays fuel into the centre of the supercharger impeller.

Ignition

Two magnetos driven at half-crankshaft speed provide the ignition for the engine. They are identical American Bosch Type 5F9LU3 or equivalent. Designated left and right when viewed from the rear of the engine, the left-hand magneto fires the rear spark plugs and the right-hand fires the front plugs on all nine cylinders. Firing order, in a clockwise direction viewed from the rear of the engine, is one, three, five, seven, nine, two, four, six and eight so that all cylinders have fired after two revolutions of the crankshaft. Spark plugs are Champion REB32E or equivalent.

Engine lubrication

This system is the full pressure, dry sump type. All moving parts of the engine are under pressure except for cylinder walls, piston pins, crankshaft roller and ball bearings which are lubricated by splash. Oil is drawn from and returned to an oil tank by a pump mounted on the engine accessory casing. The pump actually comprises two sections: the pressure section at the rear of the pump body and the scavenge section at the front; both pumps are of large tooth, spur gear design. After leaving the pressure side of the pump, oil enters the Cuno automatic filter, consisting of a cartridge of closely spaced discs separated by cleaner blades. The discs rotate at 1–2rpm, powered by the system oil pressure. This rotation imparts a combing action on the oil, helping to clean it. Once the oil has passed through the various oil passageways to lubricate the engine it drains into the sump located between cylinders five and six. This contains a mesh screen to additionally filter out foreign particles. After passing through the mesh, oil is drawn out by the scavenge pump and returned via the oil cooler to the supply tank.

Engine breathing

Normal pressure built up in the crankcase of the engine is vented out via passages in the supercharger front housing and discharged to the atmosphere through an outlet at the top of this housing. Any abnormally high pressure is vented to the atmosphere through a more direct route via a spring-loaded pressure relief valve.

Turbo supercharger

History and development

The turbo supercharger was developed by General Electric during the Twenties and Thirties. By the early Forties they were generally reliable units thanks to continual improvements in the design. An advantage over a mechanically driven supercharger is that the turbo automatically compensates for the air density reduction with altitude. As the aircraft climbs, the outside air pressure reduces, leading to a greater pressure difference between the exhaust gas pressure on one side of the turbine wheel and that on the other, at ambient pressure. The turbo therefore compensates for this drop-off in ambient pressure, increasing the

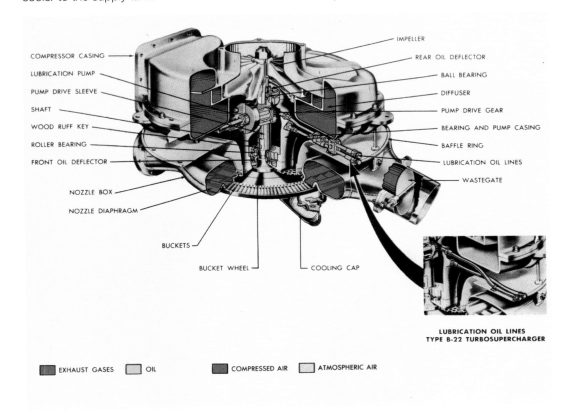

COMPRESSOR CASING
LUBRICATION PUMP
PUMP DRIVE SLEEVE
SHAFT
WOOD RUFF KEY
ROLLER BEARING
FRONT OIL DEFLECTOR
NOZZLE BOX
NOZZLE DIAPHRAGM
BUCKETS
BUCKET WHEEL
COOLING CAP

IMPELLER
REAR OIL DEFLECTOR
BALL BEARING
DIFFUSER
PUMP DRIVE GEAR
BEARING AND PUMP CASING
BAFFLE RING
LUBRICATION OIL LINES
WASTEGATE

LUBRICATION OIL LINES
TYPE B-22 TURBOSUPERCHARGER

EXHAUST GASES   OIL   COMPRESSED AIR   ATMOSPHERIC AIR

LEFT Cutaway drawing of the GE turbo, the source of power which allowed the B-17 to fly at 30,000ft. *(USAF)*

turbine speed and maintaining the power of the engine up to a high altitude.

The two models of turbo fitted to the B-17 both made by General Electric, the B-2 and the B-22, differ only in minor aspects, with the B-22 having a higher maximum rpm. Two very different control systems were fitted, however. The early type was a mechanical/hydraulic system, which required constant manipulation of the control levers by the pilots at high altitude. This was replaced by an all-electric regulator system manufactured by Minneapolis-Honeywell and was said by many B-17 pilots to be the most worthwhile modification to have been carried out to the aircraft in terms of reducing the demands on the pilots at high altitude.

Turbo installation

The turbo is installed under the engine nacelle with the turbine wheel flush with the lower surface of the nacelle. The engine exhaust feeds into the flange of the nozzle box inlet, and exhaust gas flowing through the nozzle box is allowed to exit downwards, passing through the blades (or buckets) of the turbine wheel causing it to rotate. The amount of pressure that the

exhaust exerts is controlled by the opening of a waste gate valve at the exit of the nozzle box. The turbine is connected to the impeller rotating above it by a common shaft. Intake air ducted into the centre of the impeller is derived either from a ram inlet on the wing leading edge, or via filters in the wing. Air, compressed centrifugally in the diffuser chamber is then ducted into an intercooler, which cools down the compressed air to increase the engine efficiency and reduce the likelihood of detonation. After the intercooler, the air is fed forwards to the carburettor intake where fuel is added and the air/fuel mixture is fed to the supercharger at the rear of the engine for further compression before being distributed to the cylinders for combustion. The arrangement of the air and exhaust ducting and associated components differs between the inboard and outboard engines, owing to the need to accommodate the landing gear in the inboard nacelles.

Turbo control system

Vital to maintaining the correct power output from all the engines, the electronic control system used on later B-17s was a big

BELOW Supercharger installation for both inboard and outboard engines, showing air and exhaust ducting. *(USAF)*

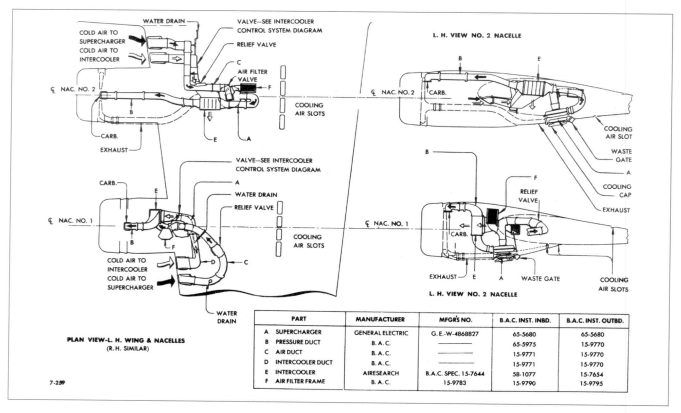

	PART	MANUFACTURER	MFGR'S NO.	B.A.C. INST. INBD.	B.A.C. INST. OUTBD.
A	SUPERCHARGER	GENERAL ELECTRIC	G. E.-W-4868827	65-5680	65-5680
B	PRESSURE DUCT	B.A.C.		65-5975	15-9770
C	AIR DUCT	B.A.C.	———	15-9771	15-9770
D	INTERCOOLER DUCT	B.A.C.		15-9771	15-9770
E	INTERCOOLER	AIRESEARCH	B.A.C. SPEC. 15-7644	58-1077	15-7654
F	AIR FILTER FRAME	B.A.C.	15-9783	15-9790	15-9795

PLAN VIEW-L. H. WING & NACELLES
(R.H. SIMILAR)

7-259

Technical specification

Wright Cyclone R-1820-97

Type: single-row air-cooled static radial engine
No. of cylinders: 9
Firing order: 1, 3, 5, 7, 9, 2, 4, 6, 8
Bore/stroke: 6.125in/6.875in
Displacement: 1,823cu in
Compression ratio: 6.7:1
Supercharger speed ratio: 7.0:1
Rated speed in rpm: 2,300
Rated bhp/rpm at sea level: 1,000/2,300
Rated bhp/rpm at 25,000ft: 1,000/2,300
Take-off bhp/rpm: 1,200/2,500
Rotation of crankshaft and propeller (from rear): clockwise
Reduction gear ratio: 16:9 (0.5625:1)
Engine weight: 1,308lb
Engine length: 48.22in
Engine diameter: 55.125in
Intake valve opens (° before TDC): 15°
Intake valve closes (° after BDC): 44°
Exhaust valve opens (° before BDC): 74°
Exhaust valve closes (° after TDC): 25°
Valve rocker clearance: 0.010in

SYMBOLS
- - - - ELECTRICAL WIRING

NOTE
SEE ELECTRICAL SECTION FOR ELECTRONIC TURBO-SUPERCHARGER CONTROL SYSTEM DETAILS.

LEGEND

A – ATMOSPHERIC & RAM PRESSURE
B – "A" PLUS TURBO PRESSURE
C – "B" MINUS PRESSURE DROP THRU INTERCOOLER
D – "C" PLUS IMPELLER PRESSURE
E – COMPRESSION
F – EXHAUST

7-262

advance over the old mechanical/hydraulic system. To control the manifold pressure, and hence the power of all four engines, the pilots operate a single knob in the cockpit. When properly calibrated, all engines produce the same manifold pressures; any small differences can be adjusted by the throttle or propeller controls. As the aircraft climbs and descends, the system compensates for changes in air density and turbo speed and adjusts the turbo waste gate to maintain the desired power setting.

The system components comprise:

- The manifold pressure selector in the cockpit, which is a dial marked from 0 to 10. A setting of 8 is used for take-off. This setting produces 1,200bhp per engine with a manifold pressure of 46in Hg when using 100 octane fuel.
- A Pressuretrol device, which is a sensitive bellows unit for measuring the induction system pressure developed by the turbo.

- A turbo governor, which registers the speed and acceleration of the turbo. This unit prevents the turbo from exceeding its 26,400rpm limit.
- The waste gate motor, which acts through a linkage arm to open and close the waste gate.
- A tube (valve) amplifier, which detects an imbalanced circuit and sends a signal to operate the waste gate motor.

The system operates at 115 volts AC, 400Hz, derived from an inverter, and from 24 volts DC from the aircraft's electrical system.

A small change in any of the parameters, such as the induction pressure, turbo speed or an adjustment of the manifold pressure selector, will cause an imbalance in the electronic circuit. This causes the amplifier to send a signal to the waste gate motor to open or close the waste gate and restore the required setting.

ABOVE It was essential that pilots and engineers understood the induction system on the B-17 because mishandling could cause the loss of an engine. This is the later version with the electronic regulator control. (USAF)

Changing an engine on the B-17

1 After disconnecting all pipe work and control cables a sling is connected to the top of the engine and the hoist prepared to lift it away from the firewall. *(All photographs Steve Carter)*

2 The old engine removed complete with mounting frame, fixed cowl and cowl flaps.

3 Only the exhaust manifold is left to remove.

4 The mount and fixed cowl removed for inspection and repair.

5 Broken Dynafocal rubber mounts must be replaced.

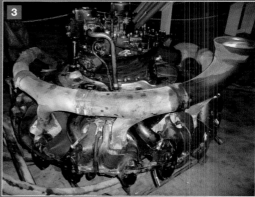

6 After crack-testing the mount is resprayed.

7 The engine mount is reunited with the cleaned fixed cowl plate.

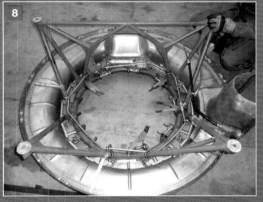

8 This view shows some of the Dynafocal mounts partly assembled to the engine mount. At the top is the cowl flap actuator.

9 All the Dynafocal mounts assembled on to the engine mount.

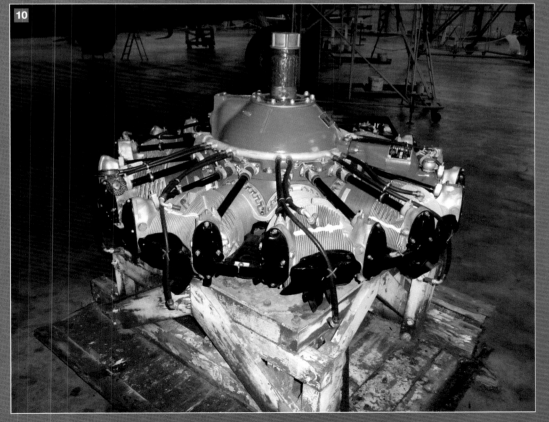

10 The reconditioned engine is removed from its crate. Replacement is a reversal of the removal procedure!

11 The exhaust system is fitted. The two black units are the magnetos.

12 Dynafocal mounts can be seen bolted to the engine. The supercharger impeller is visible through the carburettor intake.

13 The complete engine assembly fitted with cowl flaps, mounting frame, fixed cowl and intake air elbow.

14 The engine about to be lifted. The sling is attached to cylinders one and nine and the nut holding the prop shaft is released from the engine stand.

15 The engine is lifted into place.

16 It is swung back to the nacelle to line up with the mounting bolts.

17 Four bolts holding the frame to the nacelle are tightened.

18 The propeller is hoisted on to the engine.

19 Propeller installed on the refitted engine.

20 Finally, before starting the new engine the oil system is thoroughly primed to ensure oil reaches the main bearing surfaces.

113

Chapter Six

The Fort that became *Mary Alice*

The American Air Museum building at Duxford in Cambridgeshire, England, houses the Imperial War Museum's collection of American military aircraft. One of those aircraft is a B-17G, which underwent almost 20 years of restoration work that resulted in one of the most faithful and fully equipped examples of a static Fortress anywhere in the world. Her name is *Mary Alice.*

OPPOSITE Pictured in April 1993 on the hardstand at Duxford, 735 is in the markings of *Mary Alice*, the veteran B-17 from the 401st Bomb Group. *(Graeme Douglas)*

The story of the aircraft that became *Mary Alice*, like that of many of the world's surviving B-17s, began at the very end of the Second World War. Built by the Douglas Aircraft Company at Long Beach, California, and allocated USAAF serial number 44-83735, the aircraft was flown to Syracuse Air Force Base in New York to join many others straight from the factory awaiting their fate. No longer required for combat, most of the post-war B-17s ended up being scrapped even though they were brand new machines. With barely 21 hours of flight-time recorded, the aircraft was dropped from the USAAF inventory in November 1945. Fate smiled on 735 for, 18 months later, instead of being turned into aluminium ingots, it was registered as NL68269 to a company called Transocean Air Lines and converted into a luxury executive airliner for the personal use of the President of Philippine Airlines. The nose compartment was equipped as a seven-seat lounge, and the interior was fitted out with an office, complete with a refrigerator and drinks cabinet, and a toilet. The aircraft had the name *San Miguel* painted on the nose and between 1947 and 1949, with extra fuel tanks fitted in the bomb-bay, it flew regularly

ABOVE A rare colour photograph of 735 from about 1948 when the aircraft was named *San Miguel,* flying as an executive transport for the president of Philippine Airlines. *(via Bruce Orriss)*

BELOW Photographed in January 1949, *San Miguel* with civil registration NL68269. The extra windows in the waist have been added to give the passengers a better view from their luxury transport. *(C. Jannson via Scott A. Thompson collection)*

between Oakland Airport in California and Manila in the Philippines.

In October 1949, the B-17 was re-registered to an organisation known as the Assemblies of God. This Christian missionary church was intending to fly its missionaries to all parts of the world and required an aircraft for the purpose. The church had already used a converted Curtiss C-46 Commando transport aircraft for this role but now the B-17 offered increased range and better four-engine reliability. The aircraft was renamed *Ambassador II* and in the first year of operations it visited 38 countries. In an ironic twist of fate, instead of delivering bomb loads the aircraft was now delivering missionaries to spread the word of peace. It was probably the only former bomber to do so. Carrying the names of the countries it had visited emblazoned on its nose, *Ambassador II* flew to South America, Europe, Africa and Asia. After only 18 months the gospel flights stopped because in 1951 the Korean War broke out and caused difficulties for operators of former military aircraft in terms of raised insurance premiums.

Once again, 735 was sold, firstly to Leeward Aviation who then sold it on to the IGN in France in December 1952 with a total of 1,475 airframe hours. It is this organisation more than any other that has preserved so many of the world's extant B-17s, either as static or flying examples. In total from their base at Creil, near

Paris, the IGN operated a total of 13 Fortresses from 1947 through to 1989, when the last B-17 still on active duty with the IGN was unfortunately destroyed in a crash while making the feature film *Memphis Belle*. However, 735 safely ended its flying career as an aerial mapping aircraft, completing over 6,800 flying hours by February 1972, when it was eventually

ABOVE Now named *Ambassador II*, the aircraft visited many countries for the Assemblies of God organisation. This photograph dates from about 1950. *(via IWM Duxford)*

LEFT After many years of service with the IGN, during which time it bore the registration F-BDRS, the aircraft lies abandoned at Creil airfield in France without engines or undercarriage, awaiting recovery to the UK in 1975. *(Pieter Kroon, Hartsmere Logistics)*

ABOVE The aircraft is loaded on to three transport vehicles to begin its journey to Duxford. *(Pieter Kroon, Hartsmere Logistics)*

BELOW Unloading the forward fuselage section in one of Duxford's hangars. *(Pieter Kroon, Hartsmere Logistics)*

retired to become a source of spares to keep the remaining B-17s of the IGN airworthy.

Carrying the civil registration, F-BDRS, the old bomber languished in France for three years, slowly being stripped of engines, instruments, radio equipment and many other parts. The aircraft was purchased for less than £3,000 by Euroworld who brought it to the UK in 1975, together with another former IGN machine, 44-85784, which still flies today as *Sally B*. For a number of years 735 continued to be cannibalised to supply spares for *Sally B* until in 1978 the Imperial War Museum purchased it as a long-term preservation and conservation project.

The programme to restore the aircraft back to a wartime configuration began with stripping and inspecting the large sub-assemblies for corrosion. A crew of volunteer workers from the Duxford Aviation Society assisted full-time IWM engineers Ted Hagger and Eddie West in the laborious task of cleaning away the accumulations of more than 30 years of flying from inside every panel, recess and from behind every skin. This thankless job was necessary to check for corrosion and any found was treated mechanically and chemically. A number of areas required attention, including the tailplane and wheel wells, but overall the airframe was in good condition.

The next stage involved mating the sub-assemblies back together to form the recognisable shape of an aircraft. When shipped from France the wings had been split into six main pieces, the fuselage had been taken apart at the transport joint, and the tailplane and undercarriage had been removed. The aircraft was reassembled and sat back on to its wheels, but minus the outer wing panels at this stage to allow it to be moved more easily around the hangar.

Work then began to remove the many non-standard fittings both inside and outside the aircraft. With three previous operators using the aircraft for widely differing roles, most of the original military equipment had long since been removed. One of the most obvious external modifications had been the replacement of the large waist windows in the rear fuselage with a series of small rectangular passenger windows. This necessitated the re-skinning of a large area of the rear fuselage. At the same time a new rear entrance door was fabricated to replace the fold-down step which had been fitted previously. Inside the rear fuselage two rows of passenger seats and non-standard electrical equipment were removed. Throughout the interior most of the original wooden flooring was either missing or had been modified, so after the cleaning and painting of each area in turn, new floors were laid to exactly copy the originals. One problem encountered in the restoration of the radio room was that the first floor to be constructed did not extend far enough to the outside walls of the fuselage as a later photograph proved that it should. Hence, in the interests of accuracy, a second floor had to be made.

From the outset it was known that a set of gun turrets and gun mounts would be some of the most difficult military items to acquire, yet these parts were vital if the aircraft was to faithfully represent a fully equipped Air Force bomber. The museum was therefore fortunate to obtain at an early stage a ball turret from a scrapyard in the north of England; work on rebuilding this was carried on in parallel with the aircraft restoration.

The cockpit area required extensive work as

the instrument panels and electrical systems had been greatly modified. In addition, a layer of sound-absorbing material which had been applied to the cockpit walls and nose had absorbed moisture, which caused the metal

ABOVE During the early 1980s the fuselage was re-skinned to delete the passenger windows and restore the large waist apertures. New skin is painted pale yellow. (*Graeme Douglas*)

DANGER
NO ENTRY

LEFT This view shows the unrestored waist area looking forward. (*Graeme Douglas*)

ABOVE This was
the condition
of the cockpit
before restoration,
showing most of the
instruments removed
and the worn condition
of the interior. *(Graeme
Douglas)*

most of these were flown by the original pilot, Lieutenant Dan Knight and his crew. Other crews then flew the aircraft until the end of the war when it was taken back to the United States and eventually scrapped.

The aircraft was painted in an authentic olive drab and neutral grey colour scheme, and carried the squadron codes IY:G of the 615th Bomb Squadron. The restored ball turret was hung from the fuselage underside, cheek gun mounts were installed in the nose and a replica chin turret under the Plexiglass nosecone. Further work followed to add engine cowlings, the top turret dome, and a complete set of 13 replica machine guns. Now an aircraft that had always flown on the most peaceful of missions really looked like a battle veteran from the Eighth. Externally, the aircraft was practically complete.

Work did not stop at an external restoration. The crew continued to concentrate on rebuilding the interior and fitting out the fuselage to the specification of a late-production B-17G. They were determined to restore it as accurately and completely as possible. A complete set of bomb racks installed in the bomb-bay allowed replica bombs to be hung, a set of interior wooden doors and navigator and radio operator's tables were made to original drawings. Ammunition feed chutes were added to flexible guns, and the canvas bags used to catch spent cases from the waist guns were made.

A Sperry upper gun turret acquired from the US required extensive repair and restoration work before it could be fitted to the aircraft. The rare Type A-1 turret arrived with the main housing broken into three pieces and with a number of parts missing. The turret housing was repaired and the crowded interior was fitted out with the electro-hydraulic drive unit that powered the turret. Using copies of the original parts manuals and photographs, copies were made of the original missing deflector panels that guided spent cartridge cases and links out of the breeches of the guns and down to special catcher bags. A gyroscopic gun sight used to calculate deflection angles of attacking aircraft was mounted in a cradle between the guns; the cradle and linkage arms all had to be made from scratch. When complete, the heavy turret was lowered into

skin beneath to corrode. The entire cockpit was stripped, new instrument panels were made and instruments were obtained and fitted. Slowly, piece by piece, authentic military equipment began to turn up – radio sets, armour plate, gun mounts and instruments. Some items were given to the museum, others were bought from the United States, and those which couldn't be located had to be scratch-built based on information gleaned from drawings and photographs.

Acting on the advice of Eighth Air Force historian, the late Roger Freeman, the IWM decided to paint the aircraft to represent the veteran Eighth Air Force B-17, *Mary Alice,* which flew with the 615th Bomb Squadron, 401st Bomb Group, based at Deenethorpe, Northamptonshire. The original aircraft flew at least 98 missions and survived the war despite suffering damage on numerous occasions. During combat, one crew member was killed and five were wounded, one of whom was navigator Carl Hoag, who despite being partially blinded, managed to produce a flight plan to guide the pilots back after the aircraft had been damaged by flak. He received the Distinguished Service Cross for gallantry. The history of this aircraft is well documented and there are a number of photographs in existence which made the job of reproducing the external markings easier. The bomb tally and nose markings depict the aircraft as it was in the summer of 1944 after completing 30 missions;

the aircraft by crane. Once in place, access at the rear of the flight deck was very restricted, but completely authentic.

In addition to making new parts, the team paid particular attention to preserving original material whenever possible. An example can be found in the waist, where despite respraying in this area an original placard detailing the emergency procedure for dropping the ball turret in flight has been preserved. It is this attention to detail that has helped to accurately reproduce the interior of the aircraft to represent the way it left the factory in May 1945.

Mary Alice has been described as one of the most accurate and fully restored B-17s and acts as a fitting and lasting memorial to the men of the Eighth Air Force.

Mary Alice can be viewed in the American Air Museum at the Imperial War Museum, Duxford. The museum's collection of historic aircraft ranges from the earliest days of flight to modern jet fighters. The historic site, a former RAF Battle of Britain airfield, also hosts air shows which are renowned throughout the aviation world.

For more information, contact Imperial War Museum, Duxford, Cambridgeshire CB22 4QR, +44 (0) 1223 835000 or go to http://duxford.iwm.org.uk

Boeing part numbering system

oeing part numbers were evolved from drawing numbers. In the case of a typical part number, for example 9-5502-3, the first part is the drawing number – in this case 9-5502, and the last part is the dash number, -3. In an assembly made up from a number of different parts all derived from the same drawing number, each individual part carried the same drawing number but had a different dash number. Hence, it is possible that a number stamped on to a part is not the actual part number for that component, but only the drawing number of the assembly to which it is related.

Additionally, many small fasteners and other hardware items were from what is known as the standard parts series; these typically carried an 'AN' prefix (Army–Navy) followed by a series number for diameter and a dash number to denote the length, for example AN 3-6 is a bolt 3/16in diameter and 3/4in long.

LEFT The upper turret housing after repair and restoration. The turret electrical switch box is alongside.
(Graeme Douglas)

FAR LEFT The author prepares the completed turret for winching from its stand into *Mary Alice*.
(Graeme Douglas)

LEFT The turret is lifted into place.
(Graeme Douglas)

Chapter Seven

Flying the B-17

The team who maintained and flew *The Pink Lady* are acutely aware of their responsibilities in preserving this piece of living history – the only Fortress that flew in combat with the Eighth Air Force and that, until recently, was still flying. Now based in France, this unique B-17 was cared for by a small group of dedicated volunteers, mostly retired former employees of the French airline Air Inter, who gave up their spare time to work on the aircraft.

OPPOSITE Some of the ground crew and pilots in front of *The Pink Lady* after a successful display on the island of Jersey. From left to right: Jean Beaucourt, André Dominé, Michel Bézy, Roger Chauvelot, Bernard Vurpillot and Marcel Pierre. *(Gerard Boymans)*

Preparation for flight

In comparison to the preparations required to get a B-17 ready for a real combat mission, when teams of armourers, bomb loaders and ground crew often worked through the night to prepare each Fortress for a mission, a flight for an air show or training sortie today is clearly a more straightforward operation. Nevertheless, thorough preparation is essential.

To enable the aircraft to arrive at the display venue exactly at the allocated time, the aircrew work backwards from the time of their display slot, knowing how much time to allow for crew briefing, weather checks, pre-flight inspections, engine starting and warm-up, taxiing and take-off, and the flight to the destination, in order to know when to turn up at the airfield.

The ground crew arrive first to be ready to re-fuel the aircraft. They also check the oil levels and perform a general inspection of the aircraft, removing the blanking plugs from the leading edge intakes on the wings. Once the aircrew arrive the engineer in charge informs them of any snags and the captain requests how much fuel he requires. Fuel is evenly distributed between the three main tanks and the nine Tokyo tanks in each wing, so that each engine has approximately the same amount of fuel available. The most accurate way to measure fuel levels is with a marked dipstick.

A few minutes before engines are due to be started, the ground crew, having confirmed that all magneto switches are off, pull each propeller through three complete revolutions in order to clear the combustion chambers. This is an essential procedure with radial engines as oil tends to drain into the lower cylinders over time, and if the engine is started with oil still in the cylinder, then the non-compressibility of the oil can severely damage an engine. Prior to starting, a fire guard is posted with an extinguisher in case of an engine fire.

On entering the aircraft the crew carry out an interior visual inspection. They check the tail wheel oleo for correct inflation – indicated by 2⅝in clearance on the oleo strut. The tail and waist areas are checked for unsecured stores or equipment, and the control cables for damage and freedom from obstructions. In the radio compartment they confirm that the handles for emergency lowering of the landing gear and flaps

BELOW LEFT The emergency hand cranks are stowed on the left-hand side of the bulkhead. (Franck Talbot)

BELOW RIGHT *The Pink Lady* **bomb-bay also acts as a stowage area for spares and tools, which must be secured before flight.** (Franck Talbot)

are correctly stowed on the rear bulkhead; in
an emergency, these could be vital. At the rear
of the cockpit they check that the fuel transfer
pump is switched off and the transfer valves are
set to 'OFF'. They also confirm that all charts

and notes required for the flight are to hand. The
engineer in charge then signs the technical log
confirming the pre-flight inspection has been
completed, and for the aircraft's fuel state. The
captain then signs the log to accept the aircraft.

Flying a B-17

Pre-start checks
- Brakes – set
- Landing gear switch – neutral
- Generators – OFF
- Batteries/ground power – ON
- Hydraulic pressure – 450psi minimum
- Fuel transfer pump – check function
- Cowl flaps – open
- Seats – adjusted
- Trim tabs – full range and set for take-off
- VHF radio/intercom – ON
- Manifold pressure gauges – reading ambient
- Contact tower for start clearance – obtain clearance

Engine start
- Turbo boost control – set to 0
- Fuel shut-off valves – check function
- Carburettor air filters – ON
- Fire extinguisher – set to engine being started
- Throttle set – ½ to 1in open
- Mixtures – engine off or idle cut-off position
- Propellers – fully fine or high rpm
- Ignition master switch – ON
- Booster pump – ON
- Propeller – confirm clear
- Starter – energise 12 seconds then mesh
- Primer switch – ON
- Ignition – ON (after six blades)
- Mixture – rich
- Primer – OFF
- Oil pressure – rising
- Warm-up – 1,000rpm

BELOW The view from the cockpit is good for a tail wheel aircraft, but the B-17 still has to be taxied with care.
(Graeme Douglas)

Engine checks
The engines are allowed to idle at 800rpm and both magnetos on each engine are checked to see that they produce a drop in rpm when switched off in turn, indicating that the magnetos are correctly earthed. Once the oil temperatures begin to rise and the oil pressure is at least 50psi, the throttles can be opened to 1,200rpm.

Thorough warming up can take several minutes, depending on outside temperature.

Once the oil temperature has reached 70°C and the oil pressure is at a minimum of 60psi the throttles can be opened to 1,700rpm to check the CSU (Constant Speed Unit) function. The propeller pitch lever is operated through its full range of movement from fully fine to fully coarse to check that the engine rpm drops accordingly. The lever is returned to the fully up (fine) position.

Next, the feathering button for the engine being tested is pressed in. This starts the feathering operation, and the pilots observe the drop in engine rpm along with a small drop in oil pressure, before releasing the button to prevent overloading the engine. After that the throttle is set to give the static manifold pressure of about 30in Hg.

Each magneto is then checked in turn by switching off and observing the rpm drop; this should not exceed 100rpm.

Finally, all instruments related to that engine are checked. The desired readings are:

- Fuel pressure – 17psi
- Oil pressure – 70psi
- Oil temperature – 70°C
- Cylinder-head temperature – min 120°C, max 205°C
- Generator voltage – 28V

The throttle is then returned to 1,000rpm and the checks repeated on the remaining engines.

Taxiing
After unlocking the tail wheel the co-pilot releases the parking brake by applying pressure to his pedals and then pulling the brake-release knob.

As the aircraft begins to move forward both pilots test their brakes in turn and check that the hydraulic pressure is maintained between 600 and 800psi.

Instruments are checked for correct functioning and radio aids set as required. The flaps are cycled to check their operation.

Take-off checks
- Trims – set
- Throttle and propeller friction – set
- Mixtures – set
- Carb intercoolers – cold
- Propellers – fully fine
- Flaps – up or 15° as required
- Fuel shut-off valves – open
- Fuel booster pumps – ON
- Carb air filters – ON
- Altimeters – set as required
- Gyro instruments/compass – set
- Instruments – normal
- Pitot heat – as required
- Electrics – normal
- Hydraulic pressure – 600–800psi
- Harnesses and hatches – secure and closed
- Controls – full and free
- Clearance for take-off – obtained

The aircraft is taxied to the end of the runway and run forward a short distance to align the tail wheel. Then the final checks are made before take-off.

- Cowl flaps – slightly open
- Tail wheel – locked
- Fuel boost pumps – ON
- Pitot heat – as required
- Compass – check heading
- Landing lights – as required

The pilots begin the take-off run, using 2,500rpm and with a maximum manifold pressure of 39in Hg. The tail is raised at around 50mph and the aircraft lifts off at 110mph. Climbing speed is 130mph.

After-take-off checks
- Brakes – ON then OFF
- Landing gear – switch up
- Climb power – 2,300/33in Hg
- Flaps – UP
- Engine instruments – normal
- Landing gear – visual check then switch is returned to neutral
- Cowl flaps – as required
- Altimeters – as required

Pre-display checks
- Mixtures – rich
- Fuel boost pumps – ON
- Power – set to 2,300rpm/33in Hg
- Cowl flaps – as required
- Engine instruments – monitor

ABOVE Take-off is at 110mph, climb-out speed is 130mph.
(Joe Rimensberger)

Engine start sequence

The engine starting sequence recommended to pilots during the Second World War differed slightly to that used by modern operators of the B-17. The sequence illustrated is taken from the official *Erection and Maintenance Instructions* (1944) and the RAF *Pilot's Notes*. *(All photographs Graeme Douglas)*

1 The pilot checks each of the three batteries by switching on each in turn and checking that it can supply the load of the electrical inverter. When available, a ground generator is used to save draining the aircraft batteries.

2 Hydraulic pump function is checked and that the hydraulic system pressure is between 600 and 800psi.

3 The pilot unlocks the flight controls and both pilots check the full range of movement of the controls.

4 The engine cowl flaps are opened in preparation for starting.

5 The turbo supercharger control knob is set to '0' for starting.

6 Fuel shut-off valves are checked by the co-pilot to see they are 'OPEN', allowing fuel to flow to the engines.

7 The fire extinguisher is set for the engine being started.

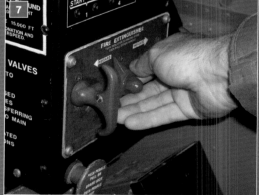

8 No 1 engine throttle lever is set approximately 1in forward.

9 Intercooler controls are set to 'COLD' by the co-pilot who then switches the carburettor air filters 'ON'.

10 The pilot sets the propeller levers to 'HIGH RPM' (fine pitch).

11 Mixture controls are set to the 'ENGINE OFF' position by the co-pilot.

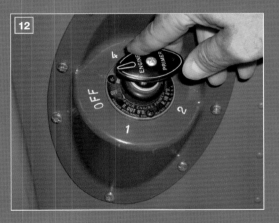

12 The engine primer is checked 'OFF' and locked by the co-pilot.

13 He then switches on the booster pump for No 3 engine, which supplies fuel to the primer.

14 The co-pilot then operates the booster pump for the engine to be started.

15 He then moves the start switch to the 'START' position and holds it there for 30 seconds.

16 While doing so he unlocks the primer and pumps it to expel air from the line.

17 The co-pilot then moves the meshing switch to the 'MESH' position while still holding down the start switch.

18 After counting six propeller blades, the pilot then turns the magneto switch to 'BOTH' for the engine being started. The co-pilot continues to operate the primer.

19 When the engine fires the pilot moves the mixture control lever to the 'AUTOMATIC RICH' position.

20 The co-pilot continues to operate the primer until the engine is running smoothly. He then returns it to the 'OFF' position.

21 The co-pilot checks the fuel pressure to ensure it is steady and within limits.

22 The fuel booster pump is then switched off.

The remaining engines are then started. The original starting sequence was 1, 2, 3, and 4; however, it is common practice today to start the inboard engines first followed by the outboards, to reduce the effect of voltage drop over the longer cable runs on the outboard starters.

In wartime *Pilot's Notes*, pilots were instructed to thoroughly check the functioning of the electrical generators and inverters followed by the manifold pressure selector. This was to test the function of the electrical turbo controls and to ensure there was sufficient boost available for take-off. This system is disconnected in many of today's airworthy B-17s, including *The Pink Lady*.

The *Pilot's Notes* cautions against excessive use of the brakes for taxiing, advising the use of power on the outboard engines in conjunction with the toe brakes.

Displaying *The Pink Lady* – Michel Bézy

ABOVE A wing-down pass along the crowd line is part of the display sequence.
(Joe Rimensberger)

The aircraft has been flown by three regular pilots: André Dominé was the original and the only pilot for ten years when the FTV purchased the B-17 from the IGN in 1987; Michel Bézy has been with the group since 1997; and Bernard Vurpillot has been flying the aircraft since 2003.

Michel Bézy, a former captain with Air France who has logged over 18,000 hours of flying time, can be accurately described as having flying in his family and his blood. He recounts how he came to be involved with The Pink Lady*:*

'I have known André Dominé since we were both following the same pilot training course at Orly in 1958. We hit it off together and have been friends ever since. We have always been aviation fanatics and I am especially interested in aircraft of the last war.

'During the early years of the B-17 operating with the FTV, I would sometimes fly with André when my long-haul duties with Air France would allow, but could only sit beside him as he was the only pilot qualified on the aircraft. I'm sure you can understand how frustrating this was for me. In 1997 the decision was taken to have a second pilot qualified and so André carried

out my qualification training during the annual Tour de France cycle race, which we followed through France from the air. I read the aircraft manual; we practised handling the aircraft with the loss of an engine and in various flight configurations. I was finally qualified to fly the B-17. I did not find the training very difficult; by today's standards the B-17 is not a complicated aircraft.'

Michel describes his impressions when taxiing the aircraft for the first time:

'When I first tried to taxi the aircraft, I didn't want to use the brakes too much for fear of overheating them, so I tried to steer using the outboard engines and with very little use of the brakes, but this was almost impossible. You have to use the toe brakes and then it is much easier. Because we operate the aircraft at light weights it is easy for it to gain speed quickly on the ground and so you have to use the brakes to prevent it running away. The maximum combat weight was about 30 tonnes; at the present time we fly at a maximum of 21 tonnes.

'The aircraft is easy to fly, it is pleasant and smooth and not too heavy, with safe handling characteristics; perhaps the hardest thing is to achieve a good three-point landing on all the

131

FLYING THE B-17

ABOVE **The B-17's ample wing area can be seen from this angle.** *(Joe Rimensberger)*

BELOW *The Pink Lady* **in its natural element flying close to La Ferté-Alais.** *(Joe Rimensberger)*

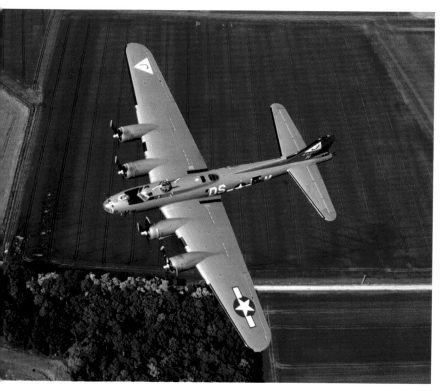

wheels together. But the pleasure is in the flying. After all it is a B-17, it is special, unique. We are very lucky to fly her. I have flown the aircraft across France at 1,500ft in unlimited visibility and witnessed extraordinary views listening to the unique sound from the four engines – that is a very special experience.

'For air shows we always try to adopt the same scenario. We try to show the aircraft from all angles, with some left- and right-hand wingovers, a high-speed flypast and a slow pass with the flaps and gear down. But we also have to be careful not to overstress the aircraft. We adopt a steep bank to show the top surface, and also we need to show the underside too. When you are manoeuvring from a 60° bank in one direction to another in the opposite direction, you have to wait for the aircraft to respond, and you are reminded that it is hard work. After a display, my flight suit is wet with sweat.

'When I flew in formation with other aircraft for the flypast over Buckingham Palace in July 2005, it was a very special event to be in the air over London at 1,000ft. It was also very hard work to keep in position in the formation. It must have been very tiring in the war for the pilots to fly such close formations for hour after hour.

'I remember once that we had some veterans visit from the States and I heard some of them talking about their pilot; they described him as "the old man" – he was 21 years old and they were 18 at the time!

'One of my best memories was flying the B-17 at an air show in Stuttgart in September 2009. We were surrounded by fighters: a Messerschmitt 109 and Focke Wulf 190 were carrying out mock attacks on us. A Spitfire, a Hurricane and a P-40 were circling around. This may have been the last time that such a collection of aircraft will fly together. It was amazing how these aircraft could turn and weave all around us. I was thinking how very difficult, if not impossible, it would be to shoot at them as they came through for a head-on pass. Then I realised how vulnerable the B-17 was in combat. They had very little chance against the enemy fighters. I understood, then, something of what the young crews in the war experienced.'

Between the three pilots there is a wealth of experience and an impressive list of aircraft types flown, as Michel recounts:

'My first ever flight was with my father, an officer in the French Air Force, who took me up in an A-24 – an Army version of the famous Douglas Dauntless, the naval bomber famed for its exploits in the Pacific War.

'I have around 18,300 hours and carried out my initial training on the Stampe biplane, then on the Morane 733 and Beech 18 followed by my time with the French Air Force on T-6, Fouga, and Noratlas. After the Air Force I undertook co-pilot training with the Air Inter Company on the Vickers Viscount, progressing to captain. I eventually got a job with Air France and started on the DC-3, then moved on to the DC-4. I gained a captain's position on the Caravelle, Boeing 707, 727, Airbus A300, A310 and finally the Boeing 747-100/200/300/400 for the last five years of my commercial career.

'André Dominé has about 17,000 flying hours, and became a civilian flying instructor after his initial training and then joined Air Inter, qualifying as captain on the Vickers Viscount, Caravelle 3, Caravelle 12, Dassault Mercure and Airbus A300.

'Bernard Vurpillot has accumulated over 14,000 flying hours, firstly as club pilot, a helicopter pilot for the Army, and then from a career of flying light public transport twin-engine aircraft, followed by work for the Port of Le Havre as a ferry pilot for tanker crews. He has extensive

ABOVE Looking over the B-17's wing while banking with a 'Little Friend' in formation. *(Graeme Douglas)*

BELOW For a number of years *The Pink Lady* has performed at the Jersey air show and is seen flying over the attractive coastline of the island. *(Gerard Boymans)*

experience flying warbirds and is qualified on the Mustang, Corsair, Skyraider and Mitchell.

'For navigation purposes we have a GPS fitted to the aircraft, which has proved very helpful in a lot of instances. When I first joined the crew we had no GPS. André, though, was very skilled at finding his way to small airfields, sometimes in bad weather, using only maps. I must admit that being used to an airline flight deck with all its instrumentation to help me, I was at first lost in *The Pink Lady* with only two VOR receivers and a radio compass to locate our position.

'The B-17 was certified for IFR flight when operating with the IGN and I and André keep our IFR qualifications current. This can be useful as we once filed an IFR flight plan in order to fly back from Jersey to Orly to avoid bad weather. As a veteran aircraft, *The Pink Lady* is only permitted to fly in VFR conditions and not in cloud or in marginal visibility.

'Sometimes it is necessary to file a flight plan, for instance when flying to a foreign country, or through certain military areas, and this is my special responsibility. The procedure has become simpler over the years; it can now be done over the phone.'

ABOVE The B-17's predictable handling makes it an excellent air show display aircraft. *(Gerard Boymans)*

RIGHT The cruising speed of the B-17 is around 150mph. *(Gerard Boymans)*

Operating limitations

The principal limitations of the B-17s flying today are governed to a large extent by legal requirements, along with the desire to treat the engines and airframe as gently as possible while still providing an exciting spectacle for air show audiences.

The legal requirements operating throughout Europe require that an ex-military type such as the Fortress has to be operated on a permit to fly basis issued by the DGAC in France (*La Direction Générale de l'Aviation Civile* – the French civil aviation authority). The aircraft is only permitted to fly in daylight in VMC conditions (Visual Meteorological Conditions), which define the weather minima (basically not in cloud and not below 5km visibility). The crew is restricted to the minimum number deemed necessary for safe operation during display flights – two pilots and two flight engineers who act as extra lookouts during displays and are ready in case of emergencies to hand-crank the landing

gear, a task requiring some considerable physical effort.

The aircraft has the following limitations placed upon it:

■ Maximum indicated airspeed (Vne) – 225mph
■ Maximum weight – 46,300lb
Maximum power settings without using turbo superchargers:
■ Take-off – 2,500rpm/39in Hg
■ Normal cruise – 1,800rpm/28in Hg
■ Max continuous cruise – 2,100rpm/31in Hg
■ Operating ceiling – 10,000ft
■ Maximum crosswind component – 16mph
In contrast, under military conditions the aircraft flew with the following limitations:
■ Maximum indicated airspeed – 305mph
■ Maximum weight – 64,500lb; maximum overload – 72,000lb
Maximum power settings using turbo superchargers:
■ Take-off – 2,500rpm/47.5in Hg
■ Max continuous cruise – 2,200rpm/36in Hg
■ Service ceiling – 35,000ft

ABOVE **On the ground at Jersey airport after the air show.** *(Gerard Boymans)*

Chapter Eight

The Engineer's View

To keep a B-17 airworthy, a great deal of work is carried out in the winter in preparation for the summer flying season. With an aircraft of this age there will inevitably be 'snags' that have occurred during the previous flying season which, although not serious, require attention. To clear the snag list and carry out the maintenance schedule, a team of volunteers are occupied one day a week for around 5 to 6 months. Additional work is often carried out by specialist contractors.

OPPOSITE Heavy maintenance such as engine changes is carried out inside the hangar. The wing is supported on jacks and trestles to allow the removal of the inboard fuel tank. *(Steve Carter)*

Maintaining the B-17

Pre-flight checks

Inspections and checks form a vital part of the maintenance procedures. Not only do they occur when the aircraft is undergoing maintenance in a hangar, but also before every flight. This is to ensure that all vital components are working correctly and have suffered no damage or failure since the last time the aircraft flew.

Before each flight, the flight crew will perform a walk around pre-flight check. This follows the procedure laid down in the wartime pilot training manual, but items relating to armaments, radio antennae and de-icer boots are omitted.

The check begins at the crew door in the waist, and proceeds around the aircraft in an anti-clockwise direction.

The pilots make sure that the ball turret, although it is a replica, is securely mounted with the guns pointing aft.

They then walk under the wing to the right main wheel. Here they examine the tyre for worn areas, cracking, and for the correct inflation. They also look at the creep marks which indicate if the tyre has slipped on the rim; check the condition of the hydraulic brake lines and look for signs of leaks; and examine the drag link and drag struts and bolts. The oleo must be correctly inflated. This is indicated by observing that there is approximately 1½in of the shiny metal exposed at the bottom of the oleo cylinder.

The interior of the nacelle is checked and examined for play in the retracting screw. A check is also made for excessive oil leaks from the tank.

Moving backward to the No 3 turbo, the turbine wheel is revolved by hand and observed for any signs of roughness or indications of cracks or damage to the blades. The condition and security of the waste gate is verified; it should be fully open. The pilots examine the exhaust system, looking for cracks or loose joints. They look at the engine cowlings for loose fasteners and correctly aligned cowl flaps, and look for signs of oil leaks from the nacelle or the engine.

Walking under the wing to No 4 engine, the inspections are replicated for this power plant.

ABOVE During the Second World War regular servicing usually took place outside. Here Pte Jack Michaelson, pictured in August 1944 has removed the main wheel from this B-17 and the wheel bearings are being re-greased. The inner brake drum is still in place on the axle. *(ww2 images)*

RIGHT The pre-flight visual inspection follows an anti-clockwise route around the aircraft. *(USAF)*

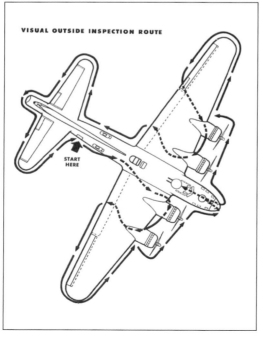

VISUAL OUTSIDE INSPECTION ROUTE

START HERE

ABOVE LEFT The ball turret (even though it is a glass fibre replica) is inspected for security. All photographs are of *The Pink Lady*. *(Franck Talbot)*

ABOVE RIGHT Correct oleo inflation is indicated by approximately 1½in of shiny metal showing at the bottom of the leg. In this case more is visible due to the aircraft being lightly loaded at the time. *(Graeme Douglas)*

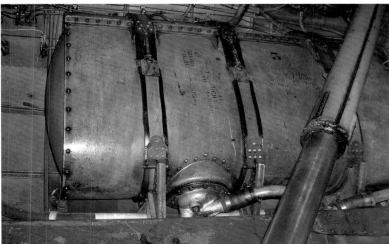

LEFT The oil tank inside the nacelle is checked for leaks and the retraction screw for excessive wear. *(Graeme Douglas)*

LEFT The turbo wheel and exhaust system are examined for cracking and loose joints. *(Franck Talbot)*

BELOW Engine cowling, exhaust shroud and cowl flaps are also inspected for security. *(Graeme Douglas)*

RIGHT The covers over the fuel filler caps are checked to see they are in place. *(Franck Talbot)*

FAR RIGHT Wing tip and the aileron condition are inspected. *(Franck Talbot)*

RIGHT The air intake covers in the leading edges of the wings must be removed before flight. *(Graeme Douglas)*

FAR RIGHT The engine and propeller are examined. *(Graeme Douglas)*

BELOW The covers must be removed from the pitot tubes. *(Graeme Douglas)*

At no time do they walk through the propeller blades.

While under the wing the pilots look for loose or missing inspection panels and signs of fuel leaks.

They examine the trailing edge of the wing and the flap for damage. The fuel filler covers are visually checked to see they are in place. The aileron fabric-covered surfaces are inspected for any tears or wrinkles. If there are external gust locks fitted to the ailerons, these must be removed.

At the wing tip the pilots ensure that the navigation light is fitted and then look along the wing leading edge, examining it for damage. The leading edge air ducts must have their blanks removed and be free of obstructions. The oil coolers are examined for leaks and damage.

Moving on to No 4 engine, the propeller blades are inspected carefully for any signs of nicks or cracks. A visual check is made of the propeller governor to see that the control cables are taut. The engine is examined for damage to the cylinders and cooling fins.

The pilots then move on to the front of No 3 engine and perform the same checks.

Moving on to the nose, they check the pitot tube, ensuring that cover has been removed.

They also examine the Plexiglass nosecone for damage.

Underneath the aircraft, the bomb-bay doors are checked to see that they are closed flush with the fuselage.

The pilots then continue to the left side of the aircraft, where they repeat the checks.

Finally, the tail assembly is inspected – the leading edge of the horizontal stabiliser for damage and the condition of the fabric on the elevators and rudder. The trim tabs are checked for alignment. Any control locks must be removed. The tail guns must be secure and the canvas cover is inspected for security.

Finally, the pilots check the tail wheel assembly. They examine the tyre for inflation, cuts, damage and wear. They also look to see that the shear bolt is intact.

Working facilities and equipment

To work on an aircraft the size of a B-17, a large amount of space in a well-lit hangar is required. In addition to the room that the aircraft itself occupies, there must be sufficient access around the airframe to allow room for working platforms or engine cranes to be manoeuvred. When large items such as engine cowlings or flying surfaces have been removed, then it is

RIGHT Working platforms and ladders are essential for working on the engines. *(Graeme Douglas)*

RIGHT These are the special tools required to remove the propeller assembly. *(Graeme Douglas)*

FAR RIGHT Sgt R. Delavance adjusts the valves on a rubber lifting bag used to raise the starboard wing of a 92nd BG, 407th BS, B-17 at Alconbury in June 1943. *(IWM D15137)*

RIGHT B-17F, 41-24502, of 306th BG, 368th BS, based at Thurleigh, Bedfordshire, undergoes repair following a crash-landing in the summer of 1942. *(IWM D11800)*

important that there is proper storage space for them on racking or shelves.

All components when removed from the aircraft must be bagged or tagged with a label detailing the item description, location, date and the signature of the engineer who removed it.

A general-purpose tool kit is needed along with a set of A/F spanners and sockets. Good-quality locking wire pliers and wire cutters are essential. The correct specialist tools for removing the propellers and an engine change kit are also required.

In order to carry out landing gear retraction checks or to lift the entire aircraft, it is necessary to support the aircraft on two hydraulic jacks of at least 13 tons capacity, and lift the tail utilising a jack capable of supporting 4 tons. The aircraft must be jacked only at special lifting cones. To remove an

FAR LEFT The rear cowling panels are easily removed as they are attached by quick-release fasteners, allowing good access to the engine controls and accessories. *(Graeme Douglas)*

LEFT The cowlings also give access to hydraulic and oil pipelines. *(Graeme Douglas)*

engine a mobile engine crane capable of lifting at least 1 ton is needed, along with a special lifting cradle that attaches to the engine. The crane is also needed to remove the propeller assembly which weighs around 380lb.

Annual servicing

Civilian-owned B-17s are today operated on what is known in the United Kingdom as a Permit to Fly. In other countries a similar system is in place. This prescribes certain restrictions on the aircraft, but in return the amount of maintenance required is in line with the limited number of hours per year that the aircraft flies. Each operator will have their own maintenance schedule agreed with the authorities in the country where the aircraft is registered. What follows is a typical example.

Maintenance schedule

This comprises inspection sheets which detail the checks and inspections required to be carried out on all parts of the aircraft. As each item on the sheet is inspected, it is signed and dated by the inspecting engineer. Completed sheets are authorised by an approved inspector or licensed engineer. If repair or rectification work is required on an item, then a separate work sheet is raised and each of the repair steps is signed off in the same way.

The first inspection interval is at 25 hours, with another inspection at 100 hours flying or after 12 months, whichever occurs first. In reality, it is almost always the annual check which occurs before 100 hours flying have been amassed.

BELOW With the cowlings removed the starter motor (shown here with the generator below) can be worked on. *(Graeme Douglas)*

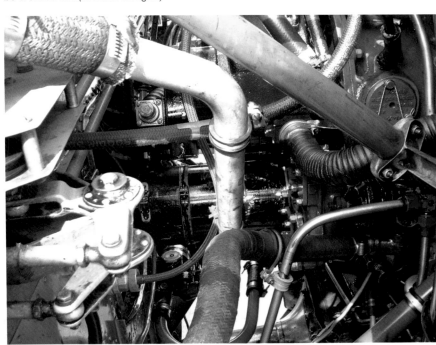

Replacing the landing gear retraction bushes

(All photographs Steve Carter unless credited otherwise.)

1 The motor at the top of the retraction screw has to be removed first. (Graeme Douglas)

2 The motor and retraction screw cover have been removed.

3 The retraction screw is then removed from the aircraft.

4 The screw has been removed with both the aircraft and the wheel jacked.

5 New bushes, including a spare. The inside thread engages with the retraction screw.

6 Both sets of screws on the bench. The unit on the right has a new bush fitted.

7 Close-up of a new bush installed in a retraction screw tube.

8 Retraction screw assembly is reinstalled. (Graeme Douglas)

9 With the aircraft still supported on jacks the landing gear is retracted to check the operation of the retraction screws.

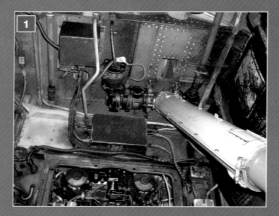

Heavy Inspections

In addition to this annual work, six items listed as Heavy Inspections are also included within the maintenance schedule. Once an item is removed from the airframe, the internal surfaces can be carefully checked for signs of corrosion or damage and the necessary repairs carried out. Before reassembly, the internal surfaces are sprayed with a corrosion inhibitor spray such as Denetrol. The Heavy Inspection items are:

1. Wing leading edges.
2. Flaps.
3. Outer wing panels.
4. Fin and Rudder.
5. Landing gear.
6. Horizontal stabiliser and elevator.

These six items are rotated into the annual schedule to allow for approximately one item per year to be included.

Out-of-Phase inspections

Other items which are covered at different inspection intervals are the Out-of-Phase inspections. These comprise items with irregular time intervals between inspections and the

Airworthiness Directives that are applicable to the aircraft:

1. Remove, strip and inspect one propeller assembly every 12 months.
2. Flexible hoses are pressure tested every 36 months, or after 72 months from new.
3. Carry out any inspections for corrosion and for internal deterioration in fabric-covered structures.
4. Carry out AD 73-20-02 and AD 77-17-11 every 12 months (bomb-bay bolts).
5. Weigh aircraft every 60 months.
6. Carry out compass swing, interval not exceeding 24 months.
7. Carry out AD 2001-22-06 every 36 months (wing spar chord bolts).

Airworthiness Directives

In order for any B-17 to continue to fly, certain Airworthiness Directives (or ADs) must be carried out at specified intervals. An AD is a notice issued by the regulatory authority, which requires that the owners or operators of the aircraft in question carry out the work listed on safety grounds in order for it to continue to fly. In the case of the B-17 there are three ADs in force, issued by the FAA in the US. All relate to inspections for corrosion and cracking in bolts and bolt holes running through the wing spars. The ADs were issued because cracks were discovered in these components in some B-17s operating in the US.

Bomb-bay

The two 12-month repetitive ADs require the removal of bolts in the bomb-bay and the inspection of the bolt holes for cracking by eddy current or borescope methods, along with a dye

Removing the fuel tank to access the wing terminal bolts

1 In order to inspect the wing terminal bolts it is necessary to remove the inboard fuel tank. Once the wing has been securely supported with jacks and trestles all the screws holding the tank access door are removed. *(All photographs Steve Carter unless credited otherwise.)*

2 Fuel and vent lines must be disconnected inside the wing.

3 The tank is then tilted through 90° and lowered on to a platform, taking care not to damage the fuel booster pump.

4 The tank is pushed away from under the wing.

5 Every 36 months the 5 numbered bolts are removed, one or two at a time, and an eddy current inspection is carried out to detect cracking of the holes. A borescope inspection is also made to check for corrosion. The procedure is repeated on each of the four wing terminals. *(Graeme Douglas)*

6 The spar chord tube end can be seen as the shiny area cleaned of paint. This must have a dye penetrant inspection to detect cracking of the tube end. Traces of the red dye are visible. *(Graeme Douglas)*

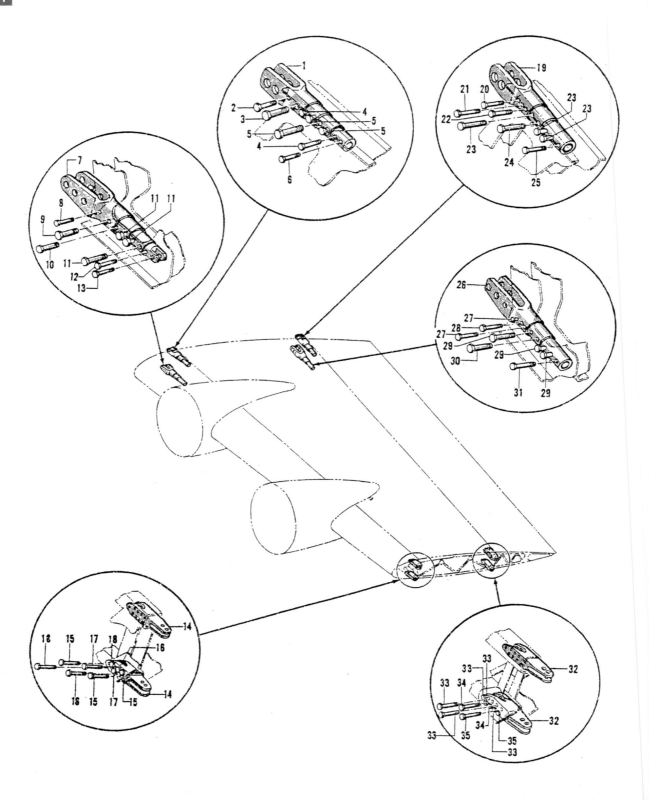

Replacing the wing spar attachment terminals on *The Pink Lady*

In order to comply with the wing spar chord AD, the wings are removed and the terminals inspected and replaced as appropriate. *(All photographs Michel Bézy unless credited otherwise.)*

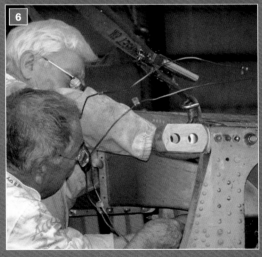

1 The wing terminals, items 1, 7, 19 and 26, and their bolts that require inspection, are shown in the illustrated parts list for the B-17. *(USAF)*

2 The right-hand wing is lifted away from the fuselage by crane.

3 Special lifting slings, manufactured to the original Boeing drawings, balance the wing weight.

4 The wing is carefully lowered on to support stands.

5 The left-hand wing is also removed.

6 Roger Chauvelot and Jacques Noble work on replacing the spar attachment terminals.

penetrant check. Providing no signs of cracking are found, the bolts can be re-fitted and the check is signed off as complete for another 12 months. If cracking is found then replacement spar caps must be fitted, or the appropriate repair must be made in accordance with the service repair manual or a repair scheme approved by the airworthiness authorities.

Wing spar chord

The AD which is repeated every 36 months requires more work, as it involves removing larger bolts that attach the terminal fitting to the wing spar chord. These bolts are located inside the wing and in order to remove them it is necessary to remove the inboard fuel tank. The bolts are then driven out and the holes inspected using the same techniques as for the smaller bolts in the bomb-bay. Again, if no cracking is found and all is well the bolts can be replaced and the inspection does not have to be repeated for another 36 months. If cracks are discovered, it is possible to repair the main spar in situ, although some wing skin would have to be removed to achieve this.

Another way to comply with the AD is to separate the wings from the fuselage and remove the wing terminals from the spar chords and carry out a full inspection. If this method is carried out, then further repeat inspections are not required.

Common faults

With an aircraft the age of the B-17, corrosion is something that must be monitored. Water ingress into panels and hollow structures will cause the aluminium surfaces to corrode over a period of time. If not treated this will damage the panel, or in the worst case it could compromise the structural integrity of the aircraft. Regular maintenance and inspections, such as ensuring that all drainage holes are kept clear, reduces the risk of water build-up. But if damage is discovered that has advanced beyond a certain stage, then the only course of action is to replace the damaged components.

The engines have a tendency to leak oil under normal conditions and occasionally also the propeller dome assembly will leak; sometimes tightening the dome lock ring will rectify this. To check the condition of the cylinders, a pressure test is performed annually, which will show if the cylinder has low compression due either to a valve problem or faulty piston rings. If the fault is related to the valves, and cannot be cured by re-seating the valve in situ, then a cylinder change is usually the quickest way to resolve the problem.

With an aircraft that is based outside during the summer months, water and moisture can cause problems. It finds its way into the ignition harnesses, which manifests itself in a large mag drop or misfiring when the engine is tested. If the aircraft is flown regularly this is not so much of a problem, as any dampness tends to get driven out.

Modern synthetic materials have tended to

RIGHT This wing tip had suffered corrosion to the upper and lower skins. Replacement aluminium is 2024 sheet. *(Graeme Douglas)*

FAR RIGHT Occasionally the propeller leaks oil, as seen here on the reduction gear casing. It can sometimes be cured by tightening the dome lock ring; if this does not work, then the seals at the blade root probably need replacing. *(Franck Talbot)*

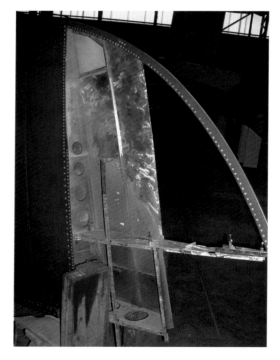

replace the original linen covering on control surfaces, largely due to their increased longevity. However, this means that the internal structures would be inspected much less frequently. It is important, therefore, that there is a regular inspection routine to check for deterioration and corrosion inside items such as elevators, ailerons and the rudder. Access panels in the fabric need to be opened and the interior of the structures must be thoroughly examined using a light source and mirror.

Fluids and lubricants

- Fuel – 100LL Avgas, 1,700 US gal contained in six main tanks. Additional 1,080gal if outer wing (Tokyo) tanks are filled.
- Oil – Aeroshell W100, 36.9gal in each of four nacelle-mounted tanks, total 148gal.
- Hydraulic fluid – Aeroshell Fluid 41, total system contents is 4gal.
- Turbo oil – Aeroshell Turbine oil 9, four tanks each of 1½gal.
- Grease – Aeroshell 22.

Operating costs

Operating an aircraft like the B-17 requires considerable financial resources. Not only is there the hourly operating cost in terms of fuel and oil consumption, but the cost of parts and engine replacement is also something to be taken into consideration when budgeting to own and operate a Fortress.

Of course, the first step for anyone wishing to own a B-17 is to find an aircraft for sale. With only 11 airworthy B-17s throughout the world, very few tend to change hands. Most of the airworthy examples have been owned by the same organisations or operators for many years and are generally not for sale. However, if someone offered the right money then it is possible that an airworthy example might be purchased for around US$1.5 to $2 million.

Another route might be to obtain a B-17 wreck, of which there are still a few around the world, and to restore it to airworthiness. This process is only for the most dedicated enthusiast, with time and money no object. At the time of writing a number of these projects are under way in the US, although it will probably be a number of years before any of the aircraft fly. It can also be argued there will be very little of the original airframe contained in what will be essentially a new-build aircraft.

The costs of maintaining a single-engine warbird such as a Spitfire are considerable, but for a four-engine heavy like the Fortress, although maintenance is not quite quadrupled, the outlay is huge. Here are some typical examples to illustrate the costs of maintenance:

The engines are normally given a TBO of 1,200hr after which a major overhaul will be required. An engine rebuild costs approximately £30,000 and it is desirable to have at least one overhauled engine in reserve. The cost of overhauling a single cylinder is around £900.

Fuel costs – at the time of writing, the average cost of 100LL Avgas is approximately £1.70 per litre in the UK, (£6.42 per US gal). In the cruise the B-17 consumes around 160 US gal/hr, equating to around £1,030/hr in fuel alone.

In Europe, after 2005 the cost of insurance rose due to EU legislation, which required the B-17 to hold the same third-party liability cover as a commercial airliner because it fell within a new weight limit group dictated by the European Parliament. The law, although aimed at commercial operators, took no account of historic 'heavies' and no exemptions were granted, forcing operators to find something in the region of an additional £25,000 per year on top of their annual insurance costs. This is one reason for the recent, but hopefully temporary, retirement of *The Pink Lady*.

An example of a 'consumable' cost item for the aircraft are the batteries: the B-17's systems are mainly electrical, including the heavy-duty motors for the landing gear and bomb-bay, and as a consequence the batteries are put under a heavy loading. The three batteries on the aircraft require replacement every few years at a cost of over £600 each.

Appendix I
Surviving B-17s

The B-17 is well represented in extant airframes when compared to other Second World War types. This is largely due to its continued use by civilian operators, especially in the firefighter tanker role, which kept the type operational through to the mid-1980s.

There are currently 48 largely complete B-17 airframes, of which 14 are airworthy or capable of flight, but only 11 regularly fly. There are also a number of wrecks which remain unrecovered and have not been included in this list.

Status Key

static = non-flyer on public display;
stored = non-flyer not on display;
potentially airworthy = maintained in a state which would require some work in order to fly, but currently no plans to do so.

Series	Serial, name(s)	Location	Status
B-17D	40-3097, *The Swoose/ Ole Betsy*	National Museum of the USAF, Wright Patterson AFB, Dayton, Ohio, USA	Under restoration to static
B-17E	41-2446, *The Swamp Ghost*	Recovered from Papua New Guinea, returned to USA, June 2010	Recovered as wreck
B-17E (XC-108)	41-2595, *Desert Rat*	Marengo, Illinois, USA	Under restoration to fly
B-17E	41-9032, *My Gal Sal*	Blue Ash Airport, Cincinnati, Ohio, USA	Under restoration to static
B-17E	41-9210	Flying Heritage Collection, Arlington, Washington, USA	Under restoration to fly
B-17F-10-BO	41-24485, *Memphis Belle*	National Museum of the USAF, Wright Patterson AFB, Dayton, Ohio, USA	Under restoration to static
B-17F-50-DL	42-3374, *Homesick Angel*	Offutt AFB, Omaha, Nebraska, USA	Static display
B17-F-70-BO	42-29782, *Boeing Bee*	Museum of Flight, Seattle, Washington, USA	Potentially airworthy
B-17G-35-BO	42-32076, *Shoo Shoo Shoo Baby*	National Museum of the USAF, Wright Patterson AFB, Dayton, Ohio, USA	Static display
B-17G-90-BO	43-38635, *Virgin's Delight*	Castle Air Museum, Atwater, California, USA	Static display
B-17G-50-DL	44-6393, *Return to Glory*	March Field Museum, Riverside, California, USA	Static display
B-17G-70-VE	44-8543, *Chuckie*	Vintage Flying Museum, Meacham Field, Fort Worth, Texas, USA	Airworthy
B-17G-85-VE	44-8846, *The Pink Lady*	Fortresse Toujours Volante, La Ferté-Alais, Cerny, France	Potentially airworthy
B-17G-85-VE	44-8889	Musée de l'Air, Le Bourget, Paris, France	Stored
B-17G-80-DL	44-83387, *Piccadilly Lily*	California Air Heritage Foundation, Los Angeles, California, USA	Fuselage only, under restoration
B-17G-85-DL	44-83512, *Heaven's Above*	USAF History Museum, Lackland AFB, San Antonio, Texas, USA	Static display

Series	Serial, name(s)	Location	Status
B-17G-85-DL	44-83514, *Sentimental Journey*	Commemorative Air Force, Falcon Field, Mesa, Arizona, USA	Airworthy
B-17G-85-DL	44-83525	Fantasy of Flight, Polk City, Florida, USA	Stored
B-17G-85-DL	44-83542, *Piccadilly Princess*	Fantasy of Flight, Polk City, Florida, USA	Static display
B-17G-85-DL	44-83546, *The Movie Memphis Belle*	Military Aircraft Restoration Corps, Chino, California, USA	Airworthy
B-17G-85-DL	44-83559, *King Bee*	SAC Museum, Lincoln, Nebraska, USA	Static display
B-17G-85-DL	44-83563, *Fuddy Duddy*	Lyon Air Museum, Santa Ana, California, USA	Airworthy
B-17G-85-DL	44-83575, *Nine O Nine*	Collings Foundation, Stow, Massachusetts, USA	Airworthy
B-17G-90-DL	44-83624, *Sleepy Time Gal*	AMC Museum, Dover AFB, Dover, Delaware, USA	Static
B-17G-90-DL	44-83663, *Short Bier*	Hill AFB Museum, Ogden, Utah, USA	Static display
B-17G-90-DL	44-83684, *Piccadilly Lilly II,*	Planes of Fame Air Museum, Chino, California, USA	Under restoration to fly
B-17G-95-DL	44-83690, *Miss Liberty Belle*	Grissom Air Museum, Peru, Indiana, USA	Static display
B-17G-95-DL	44-83718	Museu Aeroespacial, Rio de Janeiro, Brazil	Stored, dismantled
B-17G-95-DL	44-83735, *Mary Alice*	American Air Museum, Duxford, Cambridge, UK	Static display
B-17G-95-DL	44-83790	Don Brooks, Douglas, Georgia, USA	Under restoration to fly
B-17G-95-DL	44-83814, *City of Savannah*	Mighty Eighth Air Force Museum, Savannah, Georgia, USA	Under restoration to static
B-17G-95-DL	44-83863	Armament Museum, Elgin AFB, Valparaiso, Florida, USA	Static display
B-17G-95-DL	44-83868	Bomber Command Museum, Hendon, London, UK	Static display
B-17G-95- DL	44-83872, *Texas Raiders*	Commemorative Air Force, Gulf Coast Wing, Spring, Texas, USA	Airworthy
B-17G-95-DL	44-83884, *Miss Liberty*	Eighth Air Force Museum, Barksdale AFB, Bossier City, Louisiana, USA	Static display
B-17G-95-VE	44-85531*, *Shady Lady*	Evergreen Aviation Museum, McMinnville, Oregon, USA	Static display
B-17G-95-VE	44-85583	Base Aérea do Recife, Brazil	Static display
B-17G-100-VE	44-85599, *The Reluctant Dragon*	Dyess AFB, Abilene, Texas, USA	Static display
B-17G-105-VE	44-85718, *Thunderbird,*	Lone Star Flight Museum, Galveston, Texas, USA	Airworthy
B-17G-105-VE	44-85734, *Liberty Belle,*	Liberty Foundation, Douglas, Georgia, USA	Airworthy
B-17G-105-VE	44-85738, *Preston's Pride*	Highway 99, Tulare, California, USA	Static display
B-17-G-105-VE	44-85740, *Aluminum Overcast*	Experimental Aircraft Association, Oshkosh, Wisconsin, USA	Airworthy
B-17G-105-VE	44-85778, *Miss Angela*	Palm Springs Air Museum, Palm Springs, California, USA	Potentially airworthy
B-17G-105-VE	44-85784, *Sally B*	B-17 Preservation, Duxford, Cambridge, UK	Airworthy
B-17G-105-VE	44-85790, *Lacey Lady*	Adjacent to restaurant, Milwaukie, Oregon, USA	Static display minus nose section
B-17G-110-VE	44-85813, *Champaign Lady*	B-17 Project, Grimes Airport, Urbana, Ohio, USA	Under restoration to fly
B-17G-110-VE	44-85828, *I'll Be Around*	Pima Air Museum, Tucson, Arizona, USA	Static display
B-17G-110-VE	44-85829, *Yankee Lady*	Yankee Air Force, Ypsilanti, Michigan, USA	Airworthy

* Original serial number; since changed to 44-83785 and currently displayed with this number.

Appendix II
Conversion factors

The B-17 was designed and built using Imperial dimensions and units; there are differences between some US and British Imperial measures. To assist readers, the following conversion factors convert between the two systems and also metric (SI) units.

Length
1in = 25.4mm
1ft = 0.304m
1yd = 3ft = 0.9144m
1 mile = 5,280ft = 1,760yd = 1,609.3m = 0.868 nautical mile

Area
$1in^2$ = $6.45cm^2$
$1ft^2$ = $929cm^2$

Volume
$1in^3$ = $16.38cm^3$
$1,000cm^3$ = 1 litre
1 US gal = 0.83 Imperial gal = 3.78 litre
(US gallons are used throughout this book)

Weight
1lb = 0.453kg
1 short ton (US) = 2,000lb = 907kg
1 long ton (British) = 2,240lb = 1,016kg
1 metric ton (tonne) = 2,204lb = 1,000kg

Speed
1mph = 1.609kph = 0.868kt (knot = nautical mile per hour)

Pressure
1psi = $0.07kg/cm^2$
1in Hg (inch of mercury) = 0.491psi = 33.8mbar (millibar)
Atmospheric pressure at sea level is given as 29.92in Hg = 1013mbar

Power
1hp = 745.7 watts = 0.986 metric hp (PS)

Temperature
To convert temperature in °C to °F multiply the figure by 1.8 and add 32, to convert from °F to °C, subtract 32 from the figure and multiply by 0.55.

Currency
At the time of writing (September 2010) £1 Sterling is equal to US$1.59

Appendix III

American to British nomenclature

Despite supposedly sharing a common language, American and British English has a surprising number of different aeronautical terms, the most common of which are listed below. Throughout the book the terms used in original American technical documents have been used but with the spelling Anglicised.

American	British	American	British
Airplane	Aircraft	Left	Port
Antenna	Aerial	Manifold pressure	Boost
Bombardier	Bomb-aimer	Panel, outboard	Outer plane
Check valve	Non-return valve	Plexiglass	Perspex, acrylic
Co-pilot	Second pilot	Reticle	Graticule
Cylinder (hydraulic		Right	Starboard
actuator)	Jack	Screen	Filter
Dump	Jettison	Stabiliser (horizontal)	Tail plane
Empennage	Tail unit	Stabiliser (vertical)	Fin
Flight indicator	Artificial horizon	Stack	Manifold (inlet or
Gasoline (gas)	Fuel or petrol		exhaust)
Glass, bullet proof	Armoured glass	Tube (radio)	Valve
Gross weight	All-up weight	Turn indicator	Direction indicator
Ground (electrical)	Earth	Valve (fuel)	Cock
Gyro horizon	Artificial horizon	Weight empty	Tare
Land (to)	Alight	Wing	Main plane
Landing gear	Undercarriage	Wrench	Spanner
Lean	Weak		

LEFT **B-17G,** *Sentimental Journey,* **is operated in the USA by the Commemorative Air Force.** *(Patrick Bunce via Keith Wilson)*

Glossary & abbreviations

ADF – Automatic direction finding, a receiver that can be tuned to a ground beacon to locate the aircraft's bearing from the beacon.

AFCE – Automatic Flight Control Equipment.

Azimuth – Angular measurement used in military and navigational systems.

BG – Bomb Group, usually made up of 4 squadrons.

Borescope – A device for inspecting internal structures, usually consisting of a rigid or flexible wand containing optic fibres and an eye piece.

BS – Bomb Squadron, consisting of about 12 aircraft.

BTO – Bombing through overcast.

C of G – Centre of Gravity.

Chaff – Strips of aluminium, dispensed from bombers to 'clutter' enemy radar.

CSU – Constant Speed Unit.

Cupola – a defensive gun position in the shape of a dome.

DGAC – *La Direction Générale de l'Aviation Civile*, French civil aviation authority.

Duralumin – Commonly used name for type of aluminium alloy used in aircraft construction.

Dynamotor – An electric motor that is coupled to a generator, which in turn creates a high voltage output, often used to power radio equipment.

Flak – *Fliegerabwehrkanone*, German term for anti-aircraft gun.

Flight plan – Document that has to be filed before undertaking international flights or flights over water.

FTV – *Forteresse Toujours Volante* (Fortress always flying).

Geschwader – Luftwaffe fighting unit, equivalent to a squadron.

GPS – Global Positioning System.

HF – High Frequency.

H2X – Airborne ground-scanning radar for bombing through cloud, also known as *Mickey*.

Hz – Hertz, unit measuring frequency of alternating current, formerly known as cycles.

IFF – Identification Friend or Foe, transponder unit on aircraft which sent signal to Allied radar to identify it as friendly.

IFR – Instrument Flight Rules.

IGN – *Institut Géographique National* (French national geographic institute).

IMC – Instrument Meteorological Conditions.

IP – Initial Point (on bomb-run).

MAC – Mean Aerodynamic Chord.

Manifold pressure – The pressure of the mixture distributed to the cylinders, and thus an indication of the power developed by the engine. Measured in inches of mercury.

NDT – Non-Destructive Testing.

Oleo – Strut partly filled with oil and compressed gas that acts as a suspension and damping device to absorb the impact of landing.

Pathfinder – Aircraft equipped with special radar to lead the main force to the target.

PDI – Pilot Director Indicator, instrument showing that the aircraft is properly trimmed when the auto-pilot is engaged, or when there is a course deviation when on the bomb-run.

QFE – Atmospheric pressure reading set on altimeter to give height above airfield.

QNH – Altimeter set to give altitude above sea level.

SCR – Signal Corps Radio.

TBO – Time Between Overhaul (for an engine).

USAAC – United States Army Air Corps, predecessor to the USAAF.

USAAF – United States Army Air Forces, which became the current USAF in 1947.

VFR – Visual Flight Rules.

VHF – Very High Frequency.

VMC – Visual Meteorological Conditions.

VOR – VHF Omni-directional Range. A navigational aid.

Useful addresses

Airmotive Inc.
290 Airport Road
Clinton Municipal Airport
Clinton
Arkansas 72027
USA
Tel +001 501-745-4437
Rebuild specialists for Wright Cyclone
engines.

Baptiste Salis
Aérodrome de Cerny
91590 La Ferté-Alais
France
+33 (0)164 575 585
Engine overhaul specialists.

Blakey Engine Service
1486 Airway Blvd
Roanoke
Texas 76262
USA
Tel +001 817-490-1820
Re-build Wright Cyclone engines.

CFS Aeroproducts Ltd
The Alvis Works
Bubbenhall Road
Baginton
Coventry, Warwickshire
CV8 3BB
UK
Tel +0044 (0)24 7630 5873
Spare parts for Hamilton Standard
propellers.

C+S Propeller Service Inc.
4667 San Fernando Road
Glendale
California 91204
USA
Tel +001 323-245-1271
Overhaul Hamilton Standard propellers.

Deltair Airmotive Ltd
17 Aston Road
Waterlooville
Hampshire PO7 7XG
UK
+44 (0)2392 255255
Engine and propeller overhaul specialists.

Eastern Sailplanes Ltd
Peter Johnson
The Blister Hangar
Lavenham Lodge Farm
Alpheton
Suffolk CO10 9BT
UK
Tel +44 (0)1284 827300
Specialist in fabric re-covering work for control
surfaces and painting.

Kearsley Airways Ltd
Romeera House
Stansted Airport
Stansted
Essex CM24 1QL
UK
Tel +44 (0)1279 871000
Suppliers and repairers of electrical and
hydraulic components and repairs to
instruments.

Index